THE ORIGIN OF MASS

THE ORIGIN OF MASS

Elementary Particles and Fundamental Symmetries

John Iliopoulos

Great Clarendon Street, Oxford, OX2 6DP,
United Kingdom

Oxford University Press is a department of the University of Oxford.
It furthers the University's objective of excellence in research, scholarship,
and education by publishing worldwide. Oxford is a registered trade mark of
Oxford University Press in the UK and in certain other countries

Translation from the French language edition of: Aux origins de la masse
by Jean Iliopoulos © 2015 Editions EDP Sciences, Paris, France

First Edition published in 2017

Impression: 1

Published in the United States of America by Oxford University Press
198 Madison Avenue, New York, NY 10016, United States of America

British Library Cataloguing in Publication Data
Data available

Library of Congress Control Number: 2017932404

ISBN 978–0–19–880517–5

Printed and bound by
CPI Group (UK) Ltd, Croydon, CR0 4YY

Foreword

This book by John Iliopoulos begins with a question. Given that a large number of elementary particles had already been discovered, why did the announcement by CERN on 4 July 2012 of the discovery of a new particle become such an extraordinary worldwide media event, a situation highly unusual in elementary particle physics? The author tells us that this was not just a matter of finding a new particle, but, most likely, 'a window into one of the most extraordinary phenomena in the history of the Universe.' This extraordinary phenomenon is the mechanism which allows elementary particles to acquire a mass.

It is this phenomenon, which in the terminology of physicists is referred to as 'spontaneous symmetry breaking in the presence of local internal symmetries', that the author explains to us in simple but correct terms. He shows us how this phenomenon fits into our current understanding of elementary particle physics summarised in the Standard Model, to which he has himself made important contributions. In reading this book, we also come to understand that, after the CERN discovery, the Standard Model marks the boundary between the known and the unknown, and we learn why our current vision of this unknown sees knowledge of the 'infinitesimally small' of elementary particles as potentially containing knowledge of the 'infinitely large' of the observable Universe.

★ ★ ★ ★ ★ ★ ★★

The story told in this book forms a part of the history of the search for a rational understanding of the world. Let me briefly review this history.

Physics, as it is understood today, is an attempt to interpret a wide diversity of phenomena as particular manifestations of general laws which can be verified experimentally. This concept of a world governed by *verifiable* general laws is a relatively recent idea in human history. It originated in Europe during the Renaissance, and then underwent an extraordinarily rapid development. Its success is largely due to the universality of the revolutionary vision of Galileo (1564–1642). He was the first to introduce the principle of inertia, which states the impossibility of detecting a uniform rectilinear motion of a physical system, whether animate or inanimate, by an experiment performed within the system itself. This principle suggests that motion in a straight line at constant velocity need not be the result of any cause.

In accordance with the principle of inertia, in the late seventeenth century Newton formulated the celebrated law of the universal attraction of masses. Newton viewed the world as composed of tiny entities interacting with each other by means of forces which cause changes of velocity. These tiny 'point' entities have become what today we refer to as elementary particles. In the nineteenth century Maxwell introduced the concept of a field, which, in contrast to the tiny entities of Newton, fills an entire region of space. He formulated in these new terms the general laws of electromagnetism which govern all phenomena associated with electricity, magnetism, and light. The concepts of field and particle were unified during the first decades of the twentieth century by quantum mechanics, with particles becoming the 'quantum' constituents of fluctuating fields. In addition, at the beginning of the twentieth century, Einstein extended the principle of Galilean inertia to electromagnetism by developing the theory of special relativity, which modifies our concepts of time and space. He then generalised Newton's law of the gravitational attraction of masses to create the theory of general relativity, which opens up a path to the scientific study of the cosmological expansion of the Universe.

Therefore, owing in particular to the impressive progress made during the first half of the twentieth century, we arrived at a picture of the world in which all phenomena, from the atomic level to the limits of the observable Universe, appeared to be governed uniquely by two fundamental and known laws: the general relativity of Einstein and quantum electrodynamics, which is the transcription of Maxwell's theory of electromagnetism into quantum mechanics.

The gravitational and electromagnetic interactions are *long-range* interactions, that is, they act on objects separated by any distance. However, the discovery of subatomic structures indicated that additional fundamental interactions of *short range* exist whose effect is negligible at our scale. At the beginning of the 1960s their theoretical interpretation seemed to pose insurmountable problems. This is where the narrative of John Iliopoulos begins.

★★★★★★★★

In 1960 Nambu introduced into elementary particle theory the concept of 'spontaneous symmetry breaking'. Generalised in 1964 to 'local internal symmetries' by Brout and Englert, and independently by Higgs, this idea permitted these authors to construct a theoretical 'mechanism' which induces a transmutation of long-range interactions into short-range interactions by endowing the particles transmitting the force with a mass. More generally, this Brout–Englert–Higgs (BEH) mechanism allowed understanding of the origin of the elementary particle masses. What do these concepts mean and how does this mechanism work?

John Iliopoulos explains these concepts and then shows how the Standard Model of elementary particles is constructed starting from the BEH mechanism. He recounts the experimental verification and validation of the mechanism itself by the detection at CERN of the particle which is its essential element. His narrative leads us onward to theoretical speculations which, pertaining to as yet unexplored energies, go beyond the Standard Model.

From this perspective, the author shows how we are led to the necessity of fusing the 'infinitely large' of cosmology with the 'infinitesimally small' of elementary particles. The discovery in 1965 of the cosmic microwave background radiation indicated the existence of a hot primordial universe whose structure is becoming more and more accessible to us. On the one hand, improvements in the techniques for observing this primordial cosmic radiation, coupled with theoretical developments in cosmology, allow us to look back in time to the origin of the quantum fluctuations of this radiation at an epoch close to the birth of our Universe. On the other hand, theoretical and experimental results in elementary particle physics are now leading us to the analysis of the physics at energies corresponding to temperatures of around a million billions of degrees, and thus to the study of the primordial structure of the Universe at such temperatures, and then onward to the formulation of conjectures for even higher temperatures.

The data on the 'infinitely large' and the 'infinitesimally small' are coming to overlap each other more and more and are making the fusion of the theory in these two domains inevitable. Will this lead us all the way to an understanding of the birth of the Universe from quantum fluctuations of gravity? Unfortunately, the quantum extension of general relativity which might describe this era is, at best, embryonic, and it is too early to tell if we will be able to obtain a rational understanding of the birth of the Universe itself.

★★★★★★★★

The popularisation of science for a general audience is a difficult art. Explaining the achievements of theoretical physics, particularly those of contemporary physics, to an uninformed public faces a double barrier. The language of mathematics which furnishes the physicist with the mental shortcuts needed to express the concepts involved is inaccessible to the general public. Moreover, the concepts themselves are so remote from familiar ideas that they have no evident analogue. How does one explain the meaning of 'spontaneous symmetry breaking' or 'local internal symmetries', concepts which are essential for understanding modern elementary particle physics? While the mathematical description of these concepts is unambiguous, their approximate translation into everyday language requires going beyond habitual ways of thinking.

The difficulty is the greater the further away from the subject is the everyday experience of the reader. A casual perusal of this book, whose text is studded with a few equations, suggests that it is written mainly for students who are beginning work on scientific subjects. Undoubtedly it will be quite useful to this audience, but it can also be read profitably by a much broader public. Anyone interested in the fundamental questions which arise in the search for a rational understanding of the world will find some enlightenment here. The reader who is repelled by the presence of equations can simply ignore them and attentively read the accompanying explanations. The latter are sufficiently explicit to give a qualitative understanding of the subject, which in fact is the goal of scientific popularisation.

The author's description of the mathematical analysis in ordinary language without any loss of rigour in explaining the physical concepts is exemplary. Some effort on the part of the reader is, of course, required; however, the reader's task is aided by the fact that the author, in limiting himself to the essential, has brilliantly succeeded in summarising in a hundred or so pages a series of topics unfamiliar to a non-scientist. The thread of the discussion can be followed throughout the reading of the text without interruption by an accumulation of secondary details. The success of this little book at explaining ideas and facts so foreign to everyday experience in an attractive and accessible manner owing to its focus on the essential makes this a book of rare quality.

François Englert
Professor Emeritus,
Université Libre de Bruxelles,
Nobel Prize 2013

Acknowledgment

An earlier version of this book appeared in French in the «Editions de Physique». The author wants to thank Professor Michel Le Bellac and Ms Nicole Ribet for their friendly help and advice.

Contents

1

Introduction

On 4 July 2012 the European Organisation for Nuclear Research (CERN) announced a discovery which made headlines in the world media. Through a system of teleconferences, the announcement was made simultaneously at the Large Lecture Hall at CERN and in Australia, where the International Conference for High Energy Physics was taking place, but also directly on-line for the attention of universities and research centres world wide as well as the general public (Figure 1.1). It was the discovery of a new particle that completed a rather rich collection which physicists have accumulated over the years. In Appendix 1 we present a table which summarises our knowledge on what we call *elementary particles*. We find quite a large number of entries with exotic names, such as *quarks*, *gluons*, *intermediate vector bosons*, etc. Therefore, why all the fuss about the last one? (see Figure 1.2). Answering this question is the subject of this book. We want to show that it is not just a new particle but, probably, the trace of one of the strangest events that occurred in the early Universe: the phenomenon which allowed most elementary particles to acquire a mass.

In the course of presenting this extraordinary phenomenon, we will touch upon many subjects in physics and mathematics. The physics of elementary particles, naturally, but also that of cosmology and, to a certain extent, the physics of phase transitions. To prevent any misunderstandings, we want to stress here what this book *is not* supposed to be: it is not a book on elementary particles; it is not a book on cosmology; and it is not a book on phase transitions. It is not even a book on the discovery of this new particle. The discovery itself, which could make for a fascinating

Figure 1.1 The announcement of the discovery at CERN. In the centre on the podium we see Fabiola Gianotti, spokesperson of the ATLAS collaboration (also at the top right), Rolf-Dieter Heuer, CERN's Director General, and Joe Incandela, spokesman of the CMS collaboration (also at the top left). Bottom right : François Englert and Peter Higgs. Dr A. Hoecker, CERN.

story, will not be described. It is a book on the physical significance of the discovery. Out of all these subjects we will present only the elements necessary to understand the phenomenon of mass generation in the early Universe. We shall describe an important chapter of the physics of elementary particles without presenting the full theory of particles and their interactions. The reader who has already some familiarity with the field will find it easier to follow the arguments. To help them, but also all those who are unfamiliar with the subject, we include a rather long Appendix 1 which summarises everything one is supposed to know on the ultimate constituents of matter. The reader who already possesses this knowledge may safely ignore it, but the novice is strongly advised to read it *before* reading the main chapters of the book.

Figure 1.2 A collection of international press cuttings announcing CERN's discovery of the new boson. Dr A. Hoecker, CERN.

A second appendix attempts to explain very briefly some mathematical concepts to which we refer occasionally in the main text. They are some notions of group theory applied to the symmetry properties of a physical system. They have been included for completeness, but are not absolutely necessary for understanding the main subject.

The structure of the book is as follows. In the first chapter we present a brief history of cosmology: its origins and its evolution, together with our present ideas (some would say prejudices) regarding the history of the Universe. This choice requires an explanation. Indeed, it is not obvious guessing the relevance of cosmology in the microscopic world in general, and the recent CERN discovery in particular. Naturally, our first goal is to show that this choice is justified, but in fact we want to go further and show a profound connection between the infinitely large and the infinitely small. This connection will be developed throughout this book. We want to convince the reader that the structure of the Universe in all its immensity is due to the laws of microscopic physics, and that the Universe is the best laboratory for applying our ideas on the structure of matter.

The subsequent chapters develop the main subject. The directing principle is that of symmetry, a concept which has been fundamental in all the progress made during the past few decades in our understanding of the world. We first show the most intuitive aspects of this concept, namely the symmetries of our familiar space. We shall see in these examples that this concept brings us close to geometry in its simplest form. Gradually the symmetry concept will become more abstract together with the geometrical ideas. Geometry, in the mathematical sense, would have been the natural language for this book, but we have decided to stay away from mathematics. Therefore, our main task will be to translate mathematics into plain English. We shall pursue this more or less successfully; less being the more probable. Following a sequence of increasing abstractions we shall guide the reader towards a complex edifice called the *Standard Model*. This epitomises all our knowledge of the sub-atomic world and it is inside this framework that the significance of the last discovery will be revealed to us.

A remark, which is also a warning: the phenomena we are going to present in this book are those we study in high energy physics experiments, or observations in cosmology. They are phenomena which are governed by the laws of quantum mechanics. We shall try to describe them using analogies from classical physics which are closer to our everyday experience. These analogies are of limited validity. At best, they can capture some aspects of the phenomenon, never the real thing. The reader should not take them too literally.

A last remark concerns the system of units we are going to use. In classical physics we often choose the metre as a unit of length (m), the kilogram as unit of mass (kg), and the second as unit of time (s). This system is ill-adapted to the phenomena of relativistic quantum physics, which are characterised by two physical constants: the speed of light in vacuum c and Planck's constant h. The numerical values of these constants in our MKS (metre-kilogram-second) system are $c = 299792458\ \text{ms}^{-1}$ and $\hbar = h/2\pi = 1.054571726(47) \times 10^{-34}$ Js,[1] not very convenient numbers to carry around. We see that c has the dimensions of a velocity, $[c] = [\text{distance}][\text{time}]^{-1}$ and $\hbar$ those of an *action*, namely $[\hbar] = [\text{energy}][\text{time}]$. Therefore, we are going to choose a system of units in which these two constants are represented by dimensionless numbers equal to 1. This means that we shall measure all speeds as fractions of the speed of light and all actions as multiplets of $\hbar$. In this system the dimensions of all physical quantities are related and we find, for example,

$$[\text{distance}] = [\text{time}]$$
$$[\text{mass}] = [\text{energy}] = [\text{distance}]^{-1}$$

With this choice there remains only one unit to complete our system and we choose the *electronvolt*, written eV, as the unit of energy.[2] We shall mainly use its multiplets, the megaelectronvolt, $1\ \text{MeV} = 10^6$ eV and the gigaelectronvolt, $1\ \text{Gev} = 10^9$ eV. The previous relations show that time and distance will be measured in inverse eV and, using the numerical values of c and $\hbar$, we find the approximate relation:

$$10^{-15}\ \text{m} = [200\ \text{MeV}]^{-1} \tag{1.1}$$

We often call the distance 10^{-15} m one fermi, denoted by 1 f.

[1] The units are joule × second. $1\ \text{J} = 1\ \text{kg m}^2\text{s}^{-2}$. The number in parentheses in the value of $\hbar$ represents the experimental uncertainty. There is nothing similar in the value of c because, since 1983, this value is part of the definition of the international system of units.

[2] It is defined as the energy of an electron accelerated by a potential difference of 1 volt. $1\ \text{eV} = 1.602176565\ 10^{-19}$ J.

2

A Brief History of Cosmology

In contradistinction with astronomy, whose origins go back to the dawn of humanity, cosmology, as a natural science, is relatively young. As late as the beginning of the last century astronomers believed that, at large scale, the Universe was static.[1] No evolution was perceptible. This belief was based on the fact that the optical instruments of the time limited the observations to the immediate vicinity of our galaxy. The same way that our ancestors, who could not see much further than their horizon, thought that the Earth was flat, Einstein's contemporaries were convinced that the Universe had no history. Religious beliefs notwithstanding, the Universe appeared to have no evolution, neither beginning, nor end.

We credit Edwin Powell Hubble[2] as the first to shatter this world of eternal stillness; but the real history is more complex. It is true that when Albert Einstein, in around 1915, was looking for solutions to the equations of general relativity, following his time's

[1] This notion of 'large scale' is not well defined. Firstly, because it evolves with time: what was 'large scale' for early twentieth century astronomers is no longer the case. Secondly, because it depends on the object we are studying. Obviously, scientists did not wait until last century to understand that the solar system is not static. Already in 1755 the philosopher Immanuel Kant had published a study arguing that the solar system could have been formed out of a gas condensed under the influence of gravity. But this idea of creation following the laws of physics had not been applied to large structures, let alone to the entire Universe.

[2] Hubble is a rather strange character for a scientist of such reputation. He was born in 1889 at Marshfield, a small town in the state of Missouri in the United States. Apparently, he had been passionate about astronomy since he

prejudices, he was looking for a static solution, i.e. a solution independent of time. Such a solution would describe a Universe without evolution. In fact, in 1917 he introduced the concept of a *cosmological constant* following this motivation. An example of a discovery made for the wrong reasons.

The first voices to challenge this vision of immobility arose around 1920 on two fronts: firstly from the side of mathematics and then from that of physics.

Already in 1917 the Dutch mathematician Willem de Sitter (1872–1934) had shown that the equations of general relativity in the presence of a cosmological constant, contrary to what Einstein thought, admitted a solution describing a Universe in very fast expansion. At the time this solution was viewed as a mathematical curiosity, but today we have good reasons to believe that it describes the present state of our Universe. A little later, in 1922, another mathematician, from the Soviet Union this time, Alexander Alexandrovich Friedmann (1888–1925), studied Einstein's equations in the presence of a homogeneous and isotropic mass distribution. He also obtained non-static solutions of an expanding universe. Einstein thought at the beginning that the solution was wrong, but he was soon convinced of its validity.

The existence of solutions is one thing, but the physical reality could be different. Is the world we observe static, or not? In 1927 the Belgian physicist Georges Lemaître, a catholic priest but also

was a child, but during his early studies at the University of Chicago, he was mainly distinguished for his performances in sports. His records would have allowed him today to become a professional. He was more or less forced by his father to enter the Law School, a subject which did not interest him at all. It was only after the death of his father that he finally followed his own path and obtained his PhD in astronomy in 1917. He had to interrupt his studies for two years because of the First World War and it was in 1919 that he obtained his first position at the Mount Wilson Observatory, in California. The Director of the Observatory was the famous astronomer George Ellery Hale (1868–1938), who started the construction of the largest optical telescope of the time, with a diameter equal to 100 in. It was again Hale who later built the telescope of Mount Palomar (diameter, 200 in). As a result Hubble was able to work all his life with the world's best instruments.

an expert in general relativity, wrote the first formulation of the theory which today we call the theory of an expanding Universe.[3]

Lemaître's work was based on observational data. This was its great merit. Since the late nineteenth century, many astronomers had tried to measure the relative velocities of celestial bodies. The best data came from the American Vesto Melvin Slipher (1875–1969), of the Lowell Observatory in Arizona. How can we measure the speed of an object several light-years away from us, at a distance often poorly known? Slipher applied a spectroscopic method which has since become classic. Using fine spectroscopical analysis of the light received from a given source he observed that the frequencies corresponding to known atomic transitions were redshifted with respect to those measured on earth. This phenomenon is due to the relative speed with which the source moves away from us; the analogue of the frequency change of the sound of a police siren when the car is moving. It is the famous Doppler–Fizeau effect.[4] The result of these measurements was a kind of 'map' containing many luminous sources, not with

[3] If we consider atypical figures in science, the abbot, and later bishop, Lemaître is certainly one. He followed parallel studies in mathematical physics and theology. He too had to interrupt his studies because of the First World War, during which he served as an artillery officer. He entered the theology seminary in 1920 and was ordained as a priest in 1923. As a scientist he travelled all over the world and visited the best universities, Cambridge, Harvard, MIT, CalTech, where he met the most famous physicists, such as Einstein, Eddington, and Hubble. He studied with Eddington and obtained his PhD from MIT. All his life he insisted on keeping distance between his two specialities, theology and science and constantly opposed any attempt to connect them. After he introduced the theory known as 'the Big Bang', he wrote: 'As far as I can see, such a theory remains entirely outside any metaphysical or religious question'.

[4] The first to try this method was Christian Andreas Doppler (1803–53) himself in his 1842 article. Doppler was not an astronomer and his ideas on this subject were rather primitive. His article is an extraordinary mixture of an ingenious idea with totally wrong assumptions. The person who understood the power of the method was Armand Hippolyte Louis Fizeau (1819–96). In a conference in *la Société Philomatique de Paris* in 1848 he presented the method and insisted on the importance of the frequency shift in the spectral lines. From an observational point of view, the first to obtain reliable results was Hermann

respect to their positions in the sky, but to their velocities relative to the earth. The exact positions of each of these sources were little known and, even worse, the very existence of celestial bodies outside our own galaxy was not generally admitted.

Hubble's first great contribution was, precisely, to give a conclusive answer to this question. Using the excellent resolution of his Mount Wilson telescope, he could compare the luminosities of a large number of sources and establish unambiguously that many of them lay well outside our galaxy, and, even more so, some were forming entire galaxies. Although this ground breaking result was not immediately accepted by all astronomers, it is generally considered that 1925 marks the birth of extragalactic astronomy.

Lemaître had access to dual information: the distance of many luminous sources as well as their relative speeds with respect to the earth. In 1927 he established an astonishing correlation: the galaxies run away from each other with speeds which appear to be proportional to their respective distances.[5] The further away they are, the faster they appear to move. Lemaître had proof that the Universe was following the laws of the general theory of relativity. In 1930 he published the work for which he is best known, containing a model for the origin of the Universe, which he called *The primitive atom.* It was the first formulation of the theory which remained in the literature as *The Big Bang*. Figure 2.1 shows the founding fathers of modern cosmology.

The purpose of this book is not to present the evolution of our ideas on the Cosmos. We just want to show that, among the great scientific revolutions of the twentieth century, there was

Carl Vogel (1841–1907), Director of the Potsdam Observatory, who, in 1892, published a catalogue with the radial speeds of 51 sources.

[5] Lemaître first published this work under the title 'A homogeneous Universe with constant mass and increasing radius' (*Un univers homogène de masse constante et de rayon croissant*), in an obscure journal, *The Annals of Brussel's Scientific Society.* This is probably the reason why this law of speed–distance proportionality does not bear his name, but it is known as *Hubble's law*, who established it independently with better precision two years later, in 1929.

Figure 2.1 The 'fathers' of cosmology: Albert Einstein (1879–1955, Nobel Prize 1921); Edwin Powell Hubble (1889–1953); Georges Lemaître (1894–1966). Lemaître : Univ. C. Louvain , G. Lemaître Archives.

one which gave birth to a new science, that of *cosmology*.[6] Since the Universe evolves in time, it is natural to study this evolution, describing the history of the Universe from the most remote past, to its eventual fate in the most distant future.

The reader will certainly not be surprised to learn that this young science reached its age of maturity together with the development of high performance technical means of observation. Today these include: ground based stations (very large optical telescopes with sophisticated adaptive optics, radio-telescopes, cosmic ray observatories, ...), high altitude balloons, and, over recent years, a large arsenal of satellite and space stations. Something we must keep in mind is that, in the Cosmos, 'far' means 'old'. When we receive a letter from a friend who lives in a faraway country (this example refers to the times when people were still exchanging letters!), we do not learn his news of the day, but of the day the letter was posted. Similarly, when we observe

[6] People attribute to Merleau-Ponty the following poetic formula: '*... a genius of physics* (Einstein) *and a gigantic telescope* (Mount Wilson) *used by an astronomer of comparable status* (Hubble) *brought to natural philosophy, the first an idea* (general relativity), *the second a vision of the Universe* (the expansion), *and we do not know which one was more surprising and more thrilling.*' We have just seen that, in spite of its poetic value, the formula is a historical shortcut.

a distant galaxy we do not see its 'present' state, we see the state in which it was when the light started its long journey to reach the earth. This way, by looking further and further away, we obtain an increasingly older picture of the Universe; we follow with our very eyes the birth and evolution of the celestial bodies, the history of the Cosmos.

In the 1920s cosmology ceased to be a subject of philosophy and became part of the natural sciences; but with a fundamental difference. A traditional scientific method proceeds through three distinct steps: observation, elaboration of theoretical models and, finally, their experimental testing. Experiment is the repetition of observation under controlled conditions. It is only through experiment that a theoretical model is validated and it is only through such validation that it is promoted to a physical theory. But in cosmology this last step is missing. The experiment occurred once and we can only observe the results. Any mental representation we can build on the evolution of the Universe, even if it is corroborated by an impressive number of high precision observations, remains, from an epistemological point of view, as a purely theoretical hypothesis.

Figure 2.2 shows an artist's view of our present ideas on the evolution of the Universe. It is the theory of the *Big Bang*.[7] We often find in the popular literature rather imaginative descriptions of this theory. We say that 'big bang' means 'great explosion', a singular event which caused the creation of space and time. It was, to a certain extent, Lemaître's point of view: at the beginning there was a point; it was the only point that existed and there was nothing else outside that point. Warning! You should not imagine the point *inside* an ambient space. *The point was all the space*. And then

[7] It seems that this rather pejorative name was given to the theory by one of its opponents, the famous British astronomer and cosmologist Sir Fred Hoyle (1915–2001). Hoyle had his own theory, known as *the steady state theory*, which assumed a stationary Universe. This theory has been abandoned today because it does not explain in a natural way recent high precision measurements of the microwave background radiation (see footnote 11).

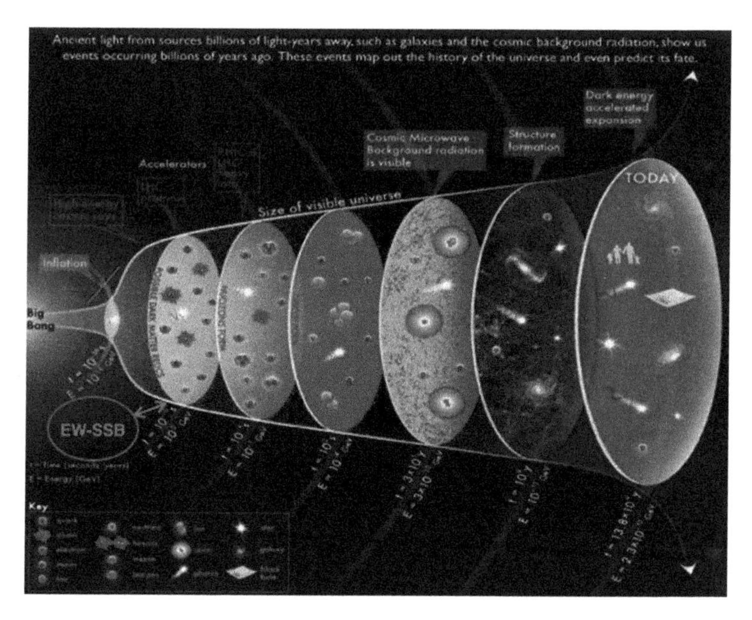

Figure 2.2 The evolution of the Universe according to the Big Bang. The axis represents the age of the Universe but also the ambient temperature expressed in eV (1 eV $\sim$ 12,000 kelvin). The numerical values are approximate and are shown only for indicative purposes. LBL, Berkeley.

the point exploded. This cosmic explosion produced an infinite amount of energy which gave rise to the entire Universe.

Although this image offers a certain approximation to this unique event, it is based simply on an extrapolation of our present knowledge, without any serious justification. We do not really know whether such an explosion did take place. Looking at Figure 2.2 we see that as we go closer to the origin, we encounter conditions characterised by extreme values of temperature (T) and matter density (D). The figure stops at a time around 10^{-43} s. Our present knowledge does not allow us to imagine the state of matter prior to this time. We think that the values of temperature and density were such that any description would require the combined effects of gravity and quantum mechanics. But, for the moment, we do not have the right equations for such a description. The extrapolation to earlier times is not justified and the 'Big Bang' is the expression of our ignorance of the state of matter under these extreme conditions.

Following Figure 2.2 we see that, as time passes, the Universe expands and cools down. We arrive soon at values of T and D accessible to our experiments. We cross 'the wall of ignorance' and enter familiar grounds.

At present this wall of ignorance stands at around 10^{-12} seconds. In the figure it is marked 'LHC'. What happened at earlier times we can only speculate. But after this time our equations are solidly anchored to experimental results. The evolution of the Universe follows precisely the laws of physics as established in our laboratories.

It is now the right moment to talk about this surprising connection between the infinitely large and the infinitesimally small we alluded to in the introduction. The study of the first is made through telescopes, that of the second through microscopes. *A priori* one wouldn't expect to find any connection between the two. Nevertheless, such a connection is not difficult to understand: the collisions between the particles we study in our accelerators reproduce, to a certain extent, a microscopic image of the conditions which prevailed in the early Universe. This is the reason why we put along the time axis in Figure 2.2 the values of the energy which are equivalent to each temperature. This way we can see how the 'wall of ignorance' moves to earlier times following the increase in the energy of our accelerators. At present the most powerful accelerator is the LHC (*large hadron collider*) of the European Organisation for Nuclear Research (*CERN*) at the border of France and Switzerland near Geneva. It is the most powerful microscope man has ever made. Its resolution power goes down to 10^{-19} m and its energy, which reaches 10^{13} eV, corresponds to a time $t = 10^{-12}$ s in Figure 2.2. From this time on, the evolution of the Universe follows our familiar laws of physics.[8]

[8] As shown in the figure, the very high energy cosmic rays go above the LHC energy, but the flux which reaches the earth is very small.

At $t < 10^{-12}$ s the values of temperature and density are very high, above anything we have studied in our laboratories. We see in the figure that, at very early times the expansion of the Universe is very fast. We call this *the inflation era.* Because of it, the entire Universe which is visible to us today, comes from a tiny region in these very early times (10^{-34} s). This is shown in the figure. Although we have no direct observation of this epoch, we have convincing indirect evidence of its occurrence. Between this time and the 'wall of knowledge' of 10^{-12} s, the Universe consists of a very hot gas of elementary particles. Among them there are certainly those we know today and whose properties are presented in Appendix 1. However, we have good reasons to believe that this primordial gas contains in addition particles which have not so far been identified. We hope that the LHC will discover them. In Chapter 4 we show that most of these particles in the primordial gas had a mass equal to zero, including those we know today, such as the electron, which appears in our experiments to be massive.

The Universe continues to expand. The temperature drops. We see in Figure 2.2 that at a temperature of the order of 300 GeV (1 GeV = 10^9 eV), we find a strange phenomenon marked as 'EW-SSB', which stands for *electroweak spontaneous symmetry breaking.* What is this? In everyday life we are used to the phenomenon of phase transitions. If we cool water we observe a sudden change: at $T = 0$ °C the water turns to ice. The water molecules are still the same, but the macroscopic properties of the system are radically different. We call this phenomenon *a phase transition* and we know many other examples of this kind. We believe that such a phase transition occurred for the entire Universe at $t \sim 10^{-11}$ s after the Big Bang, i.e. after a very short time. We are going to justify the words 'electroweak' and 'spontaneous symmetry breaking' later, but for the moment, we want to point out that during this phase transition a fraction of the energy produced by the explosion was transformed into mass, according to the well-known Einstein formula $E = mc^2$, where E is the energy, m the mass, and c the

Figure 2.3 The mass generation mechanism: Robert Brout (1928–2011). Brout died in 2011 and did not experience the triumph of the theory to the elaboration of which he had contributed; François Englert (1932–, Nobel Prize 2013); Peter Higgs (1929–, Nobel Prize 2013); Brout : F. Englert archives; Englert : F. Englert archives; Higgs : Nobel Foundation archives.

speed of light in vacuum. Most particles became massive, with the exception of the photon, which remained massless.

Fifty years ago, in 1964, three theoretical physicists, the Belgian François Englert, the American Robert Brout,[9] and Peter Higgs from Britain (Figure 2.3) proposed a mechanism which could explain this phase transition. This mechanism implied the existence of a new particle, commonly named *the Higgs particle*,[10] which has all the properties of the newly discovered particle at CERN. The *Higgshunting* lasted half a century, but it has been successful. This explains the great emotion which accompanied the discovery: it was not just a new particle, it was the element which could shed light on the mystery of mass generation in the Universe. It is the subject of this book and in the following chapters we shall guide the reader through this extraordinary adventure (see Figure 2.3).

But let us come back to Figure 2.2 and the evolution of the Universe. We see that the expansion and the resulting cooling of the Universe continues. The quarks bind among themselves and

[9] Brout spent most of his professional life in Brussels and obtained Belgian nationality.

[10] In this book we shall often call it the *Brout–Englert–Higgs or BEH particle*, which is historically more accurate.

form protons. At even lower temperature the first atomic nuclei are formed. They are those of the light elements, mainly helium. It is the epoch of *nucleosynthesis.* We are at one minute after the Big Bang.

Time goes on and the temperature drops further. Between three and four hundred thousand years later it has fallen to a level such that thermal agitation is no longer capable of keeping the electrons free. They bind to the nuclei and form the first atoms. Matter becomes electrically neutral and the photons, which interact only with electrically charged particles, can propagate freely. We call this *the moment of decoupling*, which means that the photons are no longer coupled to matter. The latter is an amorphous gas filling all space. It consists of atoms of light elements, mainly hydrogen and helium. Very slowly, under the action of gravitation, gaseous masses start concentrating. It is a very slow process; it takes several hundred million years for the first celestial bodies—we call them *proto-galaxies*—to start forming, and even more for the first sufficiently massive bodies to appear. Under their gravitational attraction nuclear reactions become possible in their interior and they start radiating. They are the first shining stars ... and there was light. The Universe starts taking its familiar form.

These sketchy notes do not pretend to give a precise account of the evolution of the Cosmos. There exist many specialised books which the reader may consult. In fact, the Universe has gone through several other phase transitions which we have not presented. Some are shown in Figure 2.2. If today we ask the question: what is the Universe made of, which are the constituents of its total energy content, we shall get a surprising answer. The visible mass, i.e. stars, galaxies, interstellar gas, everything we can see, counts for less than 5% of the total. A much larger part, of the order of 25%, seems to be composed of some sort of matter which does not interact at all with light and, therefore is not visible. We call it *dark matter* and we believe that it contains unknown neutral particles. We hope that LHC will discover them. The remaining part, of the order of 70% of the total energy balance, is not made out of any form of matter. It is a kind of diffuse energy density which manifests itself by causing an acceleration of the expansion of the Universe shown in Figure 2.2. We call it *dark energy* and its precise nature is unknown. It may be related to the cosmological constant Einstein introduced in 1917 on the basis of wrong reasoning. It is responsible for the acceleration of the expansion which has been observed in recent times and is shown in the figure.

At first sight this story resembles the scenario of a science-fiction movie and we want to show here that it is real science and not fiction. We said earlier that our views on cosmology are based on the results of observations. Their detailed presentation goes beyond the scope of this book, but we want, at least, to explain their nature.

First question: by looking at stars and galaxies, how far back can we go towards the early Universe? The answer is simple: as far back as there were luminous objects to send us light. With our present instruments we can observe the earliest proto-galaxies, the most distant celestial bodies. Figure 2.4 shows a picture taken by the satellite *Hubble*, launched in 1990. It is not obvious how to interpret this, but astronomers have been able to extract, out of the background of known stars, the signal coming from the proto-galaxies. Among its numerous instruments, *Hubble* had an ultra sensitive CCD (*charge-coupled device*) camera with a field of vision as narrow as a few fractions of a degree. It could point to star-free parts of the sky and take long-exposure pictures. With its high sensitivity it discovered these objects which are the first light sources in the history of the Universe. They are almost 13 billion years older than our time and represent the last step in the history of the Cosmos we have presented (see Figure 2.4).

Second question: can we 'see' anything older? Answer: no, if we search only for luminous objects, because before that time there were none. We should change our instruments and try to detect the diffuse radiation which existed in the Universe before the creation of any celestial body, when all matter was a hot soup of

Figure 2.4 A picture taken by the space telescope *Hubble*. It shows the 'young' Universe at age one billion years.

atoms and photons. This was at the moment of the decoupling which we placed at a few hundred thousand years after the Big Bang. These photons travel freely and we can try to detect them. Because of the Doppler–Fizeau frequency shift, we expect to see them in the domain of microwaves.

With the development of high performance detectors over past decades we have been able to obtain very precise measurements of this radiation.[11] Its fundamental importance is due to the fact that

[11] The first observation of this cosmic radiation, of the utmost importance to cosmology, was accidental. In 1964, Arno Allan Penzias (1933–, Nobel Prize 1978) and Robert Woodrow Wilson (1936–, Nobel Prize 1978), from the Bell labs,

Figure 2.5 The Universe seen by *Planck*.

these photons have remained free from interactions throughout the long evolution, from three to four hundred thousand years after the Big Bang, to our era. Consequently, they bring us some of the most ancient information we can get. With them we have a 'picture' of the Cosmos long before the formation of any celestial body, when the Universe was amorphous with no well-defined structure.

Figure 2.5 shows the latest measurements obtained by the European space mission *Planck*. Launched in May 2009 it gave us the most precise map of this radiation. The average temperature corresponds to 2.7 K or, approximately, –270 °C. It is remarkably homogeneous with variations of the order of one part in 10^5. In the picture these variations are shown using a chromatic code, red for the 'hot' regions and blue for the 'cold'. Astrophysicists

had constructed an ultra sensitive (for the time) antenna for radio-astronomy research. To their great surprise, they found that they were hindered by a diffuse background radiation which appeared to come from everywhere. They thought that it was the result of interferences from terrestrial sources and they tried, with no success, to eliminate it. It was Robert Henry Dicke (1916–97), an astronomer from Princeton, who first understood (i) that it was the cosmological microwave radiation predicted by the Big Bang theory, and (ii) that it was the most important discovery in cosmology since that of the expansion of the Universe.

are able to analyse these results and extract information on the conditions prevailing at the primordial Universe. They provide the most severe constraints on our cosmological models. In particular, all models without expansion, like Hoyle's steady state theory which we mentioned earlier, are essentially eliminated see Figure 2.5.

This is the most ancient signal that has been directly observed. As we have said already, before this era matter was ionised, electrons were not bound to protons to form atoms, and photons could not propagate freely. Thus, we cannot observe them today. For even more ancient times we have only indirect indications. For example: the number and nature of the various elementary particles we observe today, tells us something about the conditions prevailing at their creation, (the first fractions of a second after the Big Bang) or, the relative abundances of light elements, do the same for the moment of nucleosynthesis.

Could we ever observe the 'first moments' of creation? It is humanity's old dream, to see the 'beginning' of the world! This is not absolutely excluded, but we must resort to different kinds of 'messengers', other than photons. The problem is that light cannot propagate through a dense, or ionised medium. We cannot see behind a wall, the same way we cannot see the interior of a massive star, or the first moments of the Big Bang. Are there any messengers capable of doing so? We know of two, neutrinos and gravitational waves.

The neutrinos are elementary particles among those we find in Table A1.3. They are produced during nuclear reactions and they are among the first particles to be created in the early Universe. They interact very weakly with matter and they can go almost unhindered through a massive star. Using solar neutrinos we study the interior of the sun.

Gravitational waves are emitted from the acceleration of massive bodies, something analogous to the electromagnetic waves which are emitted from the acceleration of electrically charged particles. Until recently we have only had indirect evidence for their existence, through looking at binary stars, but in September 2015 the first direct observation of gravitational waves produced by the collision of two massive black holes has been reported.

In principle, either of these messengers, neutrinos or gravitational waves, could bring us direct information about the first moments of creation. Unfortunately, in both cases we must increase our detection capabilities by many orders of magnitude before we can achieve this goal. An open window on the Big Bang is still science-fiction.

3

Symmetries

In human history the concept of symmetry antedates that of science. Its importance transcends the scientific domain and extends to fields such as art and philosophy. Even a superficial account of its historical development goes beyond the scope of this book, as well as the competence of the author. We shall limit ourselves to a brief exposition of our present ideas on symmetry in microscopic physics.

In everyday language, symmetry usually invokes ideas from classical art (see Figure 3.1); likewise in science. There is an old theoretical prejudice, according to which, the 'best' theory is the most 'symmetric' one. We shall see that this prejudice has often guided us in the search for the theories of nature.

We start this chapter with an abstract idea which we shall illustrate with some simple examples. *In physics the notion of a symmetry follows from the assumption that some variable is not observable.* Therefore, no physical quantity can depend on it.

Let us consider an example: the most symmetric solid is a sphere. We can use this property to give an abstract definition of the sphere as the solid which looks the same no matter from which angle one looks at it. Consequently, the equations which define the sphere should not depend on the angles. Indeed, the equations, which take the form

$$x^2 + y^2 + z^2 = R^2, \tag{3.1}$$

are independent of the angles. It is easy to show that this property of angle independence is equivalent to equation (3.1).

Figure 3.1 The concept of symmetry is present in almost all artistic creations. However, rare are the works which present a perfect symmetry. In most cases it gives a general impression, but in the details it is broken. Here we see a reproduction of the Parthenon's western pediment.

We can generalise this idea. A physical system is symmetric if the equations that describe it remain invariant when we change the value of some variables. There is a profound theorem, proven by Amalie Emmy Noether,[1] which states that such an invariance implies a conservation law. These conserved quantities play

[1] Amalie Emmy Noether (1882–1935; Figure 3.2) is the first great female name in modern mathematics. Daughter of the German mathematician Max Noether, she studied mathematics and theoretical physics at the University of Erlangen, at a time when women did not often follow University studies. In spite of her immense talent and the famous results which bear her name, she did not have an easy life. She was invited by David Hilbert and Felix Klein to join the Mathematics Department of the University of Göttingen, probably the most famous mathematics department of the times, but her appointment was refused by the conservative majority of the Faculty who could not admit a woman to the rank of Professor. She later obtained tenure at Göttingen where she stayed until 1933, when she was forced to leave because of her jewish origins. She emigrated to the United States where she died following surgery in 1935.

Figure 3.2 Amalie Emmy Noether (1882–1935).

an important role in our understanding of physical phenomena. Some examples follow.

3.1 Space symmetries

A preliminary notion. In order to specify a point in space we use a *coordinate system.* Often it is a system of three orthogonal axes, like those shown in Figure 3.4. The position of a point is given by three numbers (x, y, and z), which denote the distances between the origin of the coordinate system and the projections of the point's position on the three axes.

By 'space symmetries' we mean the changes of the coordinate system which leave the dynamical equations invariant. They are

She was 53 years old. Among her students in Göttingen we find well-known mathematicians, such as the Dutchman, Bartel Leendert van der Waerden. The best scientists among her contemporaries, both mathematicians and physicists, had a great respect for Noether's work. Einstein wrote that she was '...the most creative mathematical genius produced since women had access to higher education.'

Figure 3.3 Euclid: The founder of geometry.

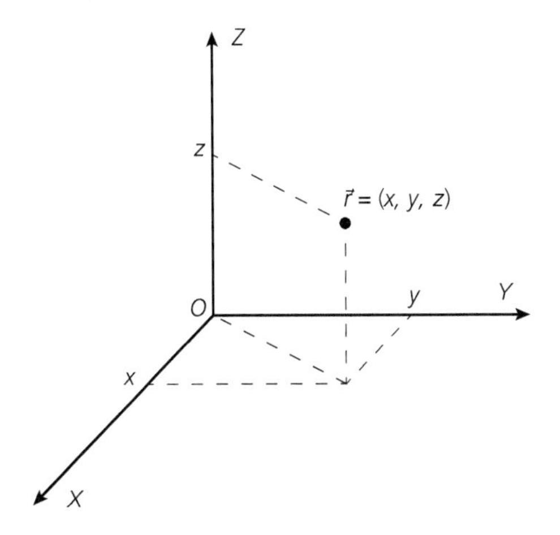

Figure 3.4 A coordinate system.

the simplest symmetries to understand intuitively. We shall present here examples of *translations* and *rotations*. We find them implicitly in the work of all geometers of antiquity, but the person who formulated them clearly was Euclid (see Figure 3.3) in his famous book *The Elements*, in which he gives the first axiomatic definition of geometry. Euclid wanted to define the notion of equality between two geometric figures and set the axiom:

> 'The superposable objects (figures) are equal'.[2]

Let us consider the example of two triangles. Euclid tells us that they are equal if we can superpose one on the other. To proceed towards this comparison we may have to perform two geometrical operations. The first is a *translation,* which means that we may have to move one of the triangles in order to bring its centre to coincide with that of the other. The second is a *rotation* in order to check whether the two are exactly superposable. Obviously, we must assume that by applying either one of these two operations, the triangle does not change. Euclid understood that this is an intrinsic property of the space, independent of the other axioms; if we want it, we must postulate it separately. We thus arrive at the notions of *translation*, or *rotation symmetry*. We shall give a concrete example presently.

3.1.1 *Translation symmetry*

Let us first consider translations. Let us assume that the space is homogeneous and the absolute position of the origin of the coordinate system is not a measurable quantity.[3] It follows that the dynamical equations should remain invariant under the transformation consisting of displacing the coordinate system by a constant vector: $\vec{x} \rightarrow \vec{x} + \vec{a}$ (Figure 3.5). We call this transformation *space translation.* We can show, using Noether's theorem, that the invariance of the dynamical equations under these translations implies the conservation of momentum.[4] In Figure 3.5 we illustrate this hypothesis for the case of a free particle. Its trajectory is a straight line represented by A. Under the effect of

[2] Και τα εφαρμόζοντα επ' άλληλα ίσα αλλήλοις εστίν.

[3] This corresponds to the intuitive idea that in a perfectly homogeneous space there is no way to specify a position in the absence of any fixed object.

[4] In classical physics the momentum of a particle is defined as the product of its mass times its velocity.

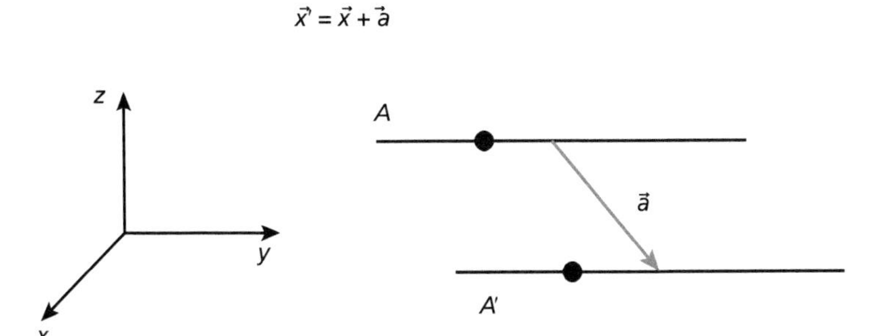

Figure 3.5 By space translation the image of a free particle trajectory A is the straight line A'.

a translation by a constant vector $\vec{a}$, the line A becomes A'. The symmetry tells us that A' is also the trajectory of a free particle.

3.1.2 *Rotation symmetry*

The same reasoning can be applied to the symmetry under rotations. The physical assumption is that space is isotropic and has no privileged direction. As a result, the particular orientation of the coordinate system is not measurable and we can perform rotations which will not affect any physical quantity. We say that the latter are *rotationally invariant*. Noether's theorem implies that to this invariance corresponds the conservation of angular momentum.

A remark is necessary here. It must be clear that the hypothesis of space homogeneity, or isotropy, applies to empty space. The presence of fixed bodies affects these properties. For example, near the earth, space is not translationally invariant: we do not breath as easily at the summit of Mont Blanc as we do in the plain. The same for rotations: a stone always falls vertically, which means that the presence of the earth induces a privileged direction, the one pointing to its centre.

3.1.3 *Space inversion*

Before closing this section on space symmetries we want to introduce a third transformation of the coordinate system which will be useful later: it is the *space inversion* $\vec{x} \to -\vec{x}$, or, in coordinates,

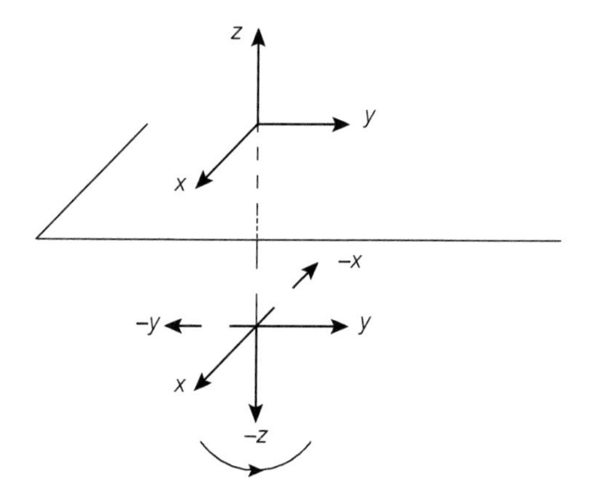

Figure 3.6 The mirror image of the coordinate system (x, y, z) is the system $(x, y, -z)$. It is obtained by an inversion followed by an 180° rotation around the z axis.

$(x, y, z) \rightarrow (-x, -y, -z)$. In physicists' jargon this transformation is often called *parity*.[5] It would have been more appropriate to call it *mirror symmetry*, because the image of the coordinate system (x, y, z), seen in a mirror placed, for example, on the (x, y) plane, is the $(x, y, -z)$, which is obtained by inversion followed by an 180 ° degree rotation around the z axis; see Figure 3.6. The same figure also shows another aspect of space inversion: it is equivalent to the transformation left ↔ right. Indeed, the mirror image of a left hand appears as a right hand. We must also stress here that a space inversion transformation cannot be reproduced by a sequence of rotations. Parity is an independent transformation.[6] The equations of classical physics, like Newton's equation for mechanics, or Maxwell's equations for electrodynamics, are

[5] A technical remark: contrary to translations, or rotations, we cannot visualise inversion as a sequence of small transformations. In mathematics we call translations or rotations *continuous transformations*, while inversion is a *discrete transformation*. This technical difference has a physical consequence: Noether's theorem does not apply and there is no associated conserved quantity.

[6] It seems that it was Lev Davidovich Landau (Nobel Prize 1962) who stated the famous aphorism: 'An acrobat may jump in the air and turn around as many times as he wants, his heart will always remain at his left!'

invariant under parity transformations, so physicists were convinced that this invariance was an exact law of nature. Therefore, it came as a great surprise when an experiment performed by Chien Shiung Wu (1912–97) of Columbia University, showed that this invariance was violated by the weak interactions of nuclear physics (see Appendix 1).

3.2 Time symmetries

As we explained previously, in order to specify a point in space we must give three numbers, its three coordinates. In order to specify an event we need four: the first three will tell us *where* and the fourth *when* the event took place. Therefore, already in classical physics, time enters as a 'fourth dimension' in our equations. Time is measured starting from an arbitrarily chosen origin. As we did for space, we assume that this choice is unimportant; physics should not depend on the way we measure time, be it since the birth of Christ, or the foundation of Rome. Consequently our equations should be invariant under *time translations* of the form: $t \rightarrow t + \tau$, where τ denotes an arbitrary time interval. Noether's theorem tells us that to this invariance corresponds the conservation of energy.

We can also consider the discrete transformation of *time reversal* $t \rightarrow -t$. Applied to a system of moving particles this transformation results in the reversal of all velocities, since the speed of a particle, which equals the distance covered in unit time, changes sign under this transformation. As for space inversion, invariance under this discrete transformation is not related to a conservation law. Classical physics is invariant under time inversion,[7] but very precise experiments had already shown, in 1964, that the

[7] This statement requires an explanation. Newton's equation, which describes the motion of a particle, is indeed invariant under $t \rightarrow -t$, but everyday experience shows that very often the evolution of a macroscopic physical system is time irreversible. Air moves from a high pressure region to one of lower pressure, but not the other way around. We get older, but not younger. The emergence of this macroscopic irreversibility out of microscopic equations

interactions among elementary particles present a slight violation of this invariance.

We have just seen the role of time as a 'fourth dimension' in classical physics. In the framework of Einstein's theory of special relativity this notion acquires a more profound significance. In a relativistic theory we can talk about the combined *space-time* which admits transformations that mix time with the space components. These transformations generalise in a non-trivial way the ones we have presented so far and are part of our microscopic theories. However, their detailed study goes beyond the scope of this book.

3.3 Internal symmetries

All the symmetries we have seen so far concern transformations of space and time. They are geometric transformations, in the real sense of the word, easy to visualise and understand intuitively. We shall need a certain degree of abstraction in order to imagine transformations which do not affect the coordinate system of space and time, but do change the dynamical variables of the problem we are studying. We shall call the resulting symmetries *internal symmetries*. A characteristic example is provided by Heisenberg's theory of *isotopic spin*, first introduced in 1932, which in today's language, can be expressed as follows.

In Appendix 1 we point out that elementary particles often possess a proper angular momentum which we call *spin*. As is the case with many quantities in quantum mechanics, spin can take only discrete values, integer, or half-integer. For example, the spin of the electrons equals 1/2. Like angular momentum, spin is a vector in space and we can look at its projection on one of the axes of the coordinate system, for example the z axis. Thus, the component s_z of the electron spin can take two values $s_z = +1/2$ and $s_z = -1/2$.

which are time reversible is a complex question which we shall not address in this book.

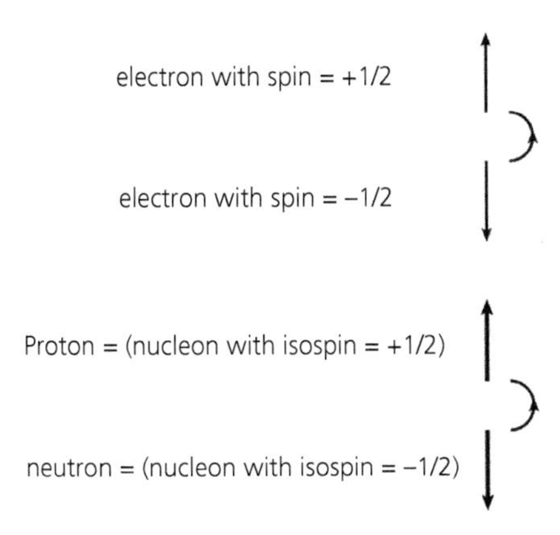

Figure 3.7 A space rotation exchanges electrons with opposite spin. A rotation in isotopic spin space exchanges protons with neutrons.

Strictly speaking we should say that we have two kinds of electrons, those whose spin points upwards and those downwards, as indicated in Figure 3.7. The reason why this distinction is useless is the existence of transformations, to wit 180° rotations around either the x or the y axis, which are symmetries of the theory and transform an electron with $s_z = +1/2$ to one with $s_z = -1/2$, and vice versa.

It is this spin analogy that Heisenberg used in his famous 1932 article on the nuclear forces. The constituents of nuclei are protons and neutrons.[8] The former are electrically charged, the latter are neutral. If we neglect this difference, the experimental results show that, as far as the nuclear forces are concerned, protons and neutrons play very similar roles: their masses differ very little and the nuclear energy levels do not change significantly if we interchange one for the other. It is this exchange symmetry that Heisenberg generalised to continuous transformations.[9] He postulated that the system proton–neutron behaves as if there were

[8] It is worth noticing that Heisenberg wrote this article in the same year that the neutron was discovered.

[9] In Heisenberg's 1932 paper the symmetry under these transformations was not complete. It was extended and completed in 1938 by Nicholas Kemmer.

only one particle, which we shall call a *nucleon,* which can appear in two states, either as a *proton,* or as a *neutron.* In this work Heisenberg introduced for the first time a physical quantity which is not connected at all with space-time. Today we call it *isotopic spin, or isospin* and we attribute to it, formally, the same properties as those of ordinary spin: it can take only integer, or half-integer values and Heisenberg assumed that the nucleon has isotopic spin equal to 1/2. Like spin, isotopic spin is assumed to be a vector. However, it is *not* a vector of ordinary space, but of another, abstract space, the *isospace.* In this abstract space we can consider a coordinate system where the component t_z of the nucleon's isospin could take two values, $t_z = +1/2$, or $t_z = -1/2$. Heisenberg identified the first state with a proton and the second with a neutron. As we did with the electron spin, we can avoid talking about two distinct particles if we assume that the theory is invariant under rotations in this new space. An 180° rotation around the x axis in isospace transforms a proton into a neutron (see Figure 3.7). This is the concept of an *internal symmetry.* The isotopic spin symmetry is approximate because the previous reasoning neglected the effects due to the electric charge of the proton. But, as far as the nuclear forces are concerned, it is a good approximation. For the first time in physics we have considered coordinate transformations other than those of our familiar space-time.

From a purely mathematical point of view, if we have a system described by dynamical equations, it is natural to associate with it a space; it is the space in which all the transformations act, leaving the equations invariant; in other words all the symmetries of the system. It follows that, for nuclear physics, the 'space' has seven dimensions, four of our familiar space and time, and three for isospace. Indeed, this last one is three-dimensional, isomorphic to our familiar space. However, the idea was later generalised to more complicated spaces, as we were discovering larger internal symmetries. The notion of space became abstract, the space of microscopic physics became a mathematical multidimensional object with a complicated topology, of which only a part, the space of our everyday experience, is accessible to our senses. We shall often use the term *internal space* to denote this abstract space in which the transformations of an internal symmetry act.

3.4 Local or gauge symmetries

With the exception of space or time inversions, all the transformations we have introduced so far depend continuously on one or more parameters. The space translations depend on the three components of the vector $\vec{a}$, those of time on τ, the rotations in either ordinary space or isospace on the three angles of rotation etc. These parameters are independent of the space-time point $(\vec{x}, t)$.

We shall now introduce a second abstraction of the concept of space by considering transformations depending on parameters which are themselves arbitrary functions of $(\vec{x}, t)$. Such transformations are called *local*, or *gauge transformations*.[10] The physical motivation to introduce this abstract notion is not obvious, but

Figure 3.8 Abstract symmetries: Werner Heisenberg (1901–76, Nobel Prize 1932); Chen Ning Yang (1922–, Nobel Prize 1957); Robert Mills (1927–99). Heisenberg : Max Planck Institute, Munich archives.

[10] The origin of the term *gauge* goes back to the work of Hermann Klaus Hugo Weyl (1885–1955) in 1918 in which he attempted to construct a theory invariant under 'scale' transformations. He wanted to have a theory in which the unit of length, 'the gauge', would change from one point of space to

it has a mathematical, or even an aesthetic origin; the equations we obtain by imposing this invariance have a richer mathematical content and a higher predictive power. It is through its consequences that the requirement for gauge invariance will be justified.

Independently of one's motivations, the first question we should ask is: can gauge transformations be symmetries of a physical theory, in other words, can they leave its dynamical equations invariant? At first sight the answer seems to be *no!*

3.4.1 Local translations

Let us look at space translations. The local ones can be described by $\vec{x} \rightarrow \vec{x} + \vec{a}(\vec{x}, t)$. If the three components of $\vec{a}(\vec{x}, t)$ are functions of the space-time point $(\vec{x}, t)$, the image of the straight line A which corresponds to the trajectory of a free particle, would be a curve A'' (see Figure 3.9). No free particle would follow such a trajectory. Indeed, the equations of motion, which in this case take

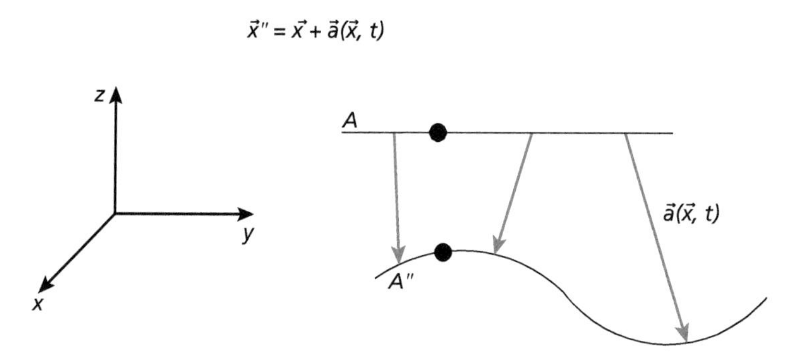

Figure 3.9 Local space translations.

another. It was natural to call such a theory *gauge invariant.* Even if this motivation was dropped later, Weyl continued to use the word 'gauge' for every transformation whose parameters depended on the space-time point and this terminology remained in the scientific literature. In this book we shall use the terms 'local' or 'gauge' transformations interchangeably.

the simple form: acceleration $= \ddot{\vec{x}} = 0$,[11] do not remain invariant under the translation $\vec{x} \rightarrow \vec{x} + \vec{a}(\vec{x}, t)$. A particle will not follow a trajectory like A'', unless it is subject to specific forces.

Can we determine these forces? In other words, can we find the dynamics which remains invariant under local translations? The question seems to be only geometric, with no obvious physical significance. Therefore we would expect to find an answer interesting only in mathematics, but not in physics. Nevertheless, the surprising result is that the dynamics we find does not describe some obscure force, but one of the fundamental forces of nature, the force of gravitation. The equations which remain invariant under local translations are the equations of general relativity. Einstein's initial reasoning, which established general relativity by trying to enlarge the applicability of the principle of equivalence, uses precisely the same concepts.

We shall not prove this result here; we shall limit ourselves to explaining an underlying physical principle. A simple dynamical postulate states that the trajectory of a free particle between two points in space is given by the shortest path that joins these points. For the example of the particles of Figure 3.5, they are the straight lines A and A'. In mathematics, the shortest path between two points is called *a geodesic* and it is a concept which characterises the geometry of space. For example, on the surface of a sphere the geodesic between two points is the arc of the circle which goes through these two points and whose centre coincides with the centre of the sphere. Let us come now to Figure 3.9. In order for a free particle to follow the curve A'', the latter should be a geodesic of the ambient space. It is precisely what general relativity predicts: the presence of massive bodies deforms the geometry of the space and straight lines are no longer geodesics. The conclusion is that *gravitation has a geometric origin.*

Is this relation between geometry and dynamics accidental? Is it limited to gravitation, or is it of more general validity?

[11] A free particle is one which is subject to no force and therefore, according to Newton's equation, its acceleration vanishes.

3.4.2 *Internal gauge symmetries*

In 1926, just two months after Erwin Schrödinger (see Figure A1.5) had published the equation which bears his name in quantum mechanics, the Russian physicist Vladimir Aleksandrovich Fock (1898–1974) made a fundamental contribution which has not been fully appreciated, even by physicists. The basic quantity which enters Schrödinger's equation for a particle is its *wave function* $\Psi(\vec{r}, t)$ (see Box 3.1). It is a complex valued function of the space point $\vec{r}$ and time t, and the square of its modulus gives the probability density of finding the particle at the point $\vec{r}$ at time t. It follows that for a single particle, only the modulus of its wave function has a physical meaning, the phase is not a measurable quantity. In mathematical terms a change of phase of the wave function is written as: $\Psi(\vec{r}, t) \to e^{i\theta}\Psi(\vec{r}, t)$, where θ is a constant phase (see Box 3.1). Under this transformation Schrödinger's equation remains invariant and, following the discussion at the beginning of this chapter, the phase must correspond to a *symmetry*. It is an internal symmetry, because the transformation does not affect the space-time coordinate system. In fact, this is the first example of an internal symmetry in quantum theory.

Box 3.1: The wave function in quantum mechanics

The history of quantum mechanics is quite complex and there exist several specialised books on the subject. Its origins go back to the end of the nineteenth century with experiments on the spectrum of black body radiation and the first results on atomic spectroscopy. The first theoretical work which broke away from classical ideas was that of Planck (see Figure A1.3) who, in 1900, postulated that the energy exchanges between matter and radiation are *quantised*, with quanta proportional to the frequency of radiation. The proportionality constant is the famous *Planck's constant, h*. In 1905 Einstein followed this principle and offered the first explanation of the photoelectric effect in quantum terms. This period ends sometime between 1913 and 1916 with the formulation by Niels Bohr (see Figure A1.3) and Arnold Sommerfeld[a] of the quantisation rules for the electronic orbits in atoms. This is the *old quantum theory.*

What was missing was a dynamical foundation. The rules were postulated but they did not follow from any precise equations. This need was fulfilled during the 1920s by Werner Heisenberg (see Figure 3.8), who proposed a system of equations in matrix form, and by Erwin Schrödinger with his differential equation. It is Dirac (see Figure A1.5) who showed that the two formulations were, in fact, equivalent.

In 1923, Louis de Broglie (see Figure A1.3) proposed a correspondence law between a particle and a wave. This is the famous particle–wave duality law. Schrödinger, probably trying to answer a question by Peter Debye,[b] made this correspondence precise by writing a wave equation for de Broglie's matter waves. This is the famous Schrödinger equation.

Consider a particle of mass m in a potential $V(\vec{r})$. Schrödinger's equation involves the wave function $\Psi(\vec{r}, t)$, which satisfies:

$$i\hbar\frac{\partial\Psi}{\partial t} = \frac{\vec{p}^2}{2m}\Psi + V(\vec{r})\Psi, \tag{3.2}$$

where $\vec{p}$, with components p_x, p_y, and p_z, is the momentum of the particle which, in Schrödinger's formulation, is given by

$$p_x = -i\hbar\frac{\partial}{\partial x} \; ; \; p_y = -i\hbar\frac{\partial}{\partial y} \; ; \; p_z = -i\hbar\frac{\partial}{\partial z}. \tag{3.3}$$

The factor $i = \sqrt{-1}$ which multiplies the left-hand side of Schrödinger's equation shows that the wave function Ψ takes

[a] Arnold Johannes Wilhelm Sommerfeld (1868–1951).

[b] Petrus Josephus Wilhelmus Debije (1884–1966, Chemistry Nobel Prize 1936).

complex values. Like every complex number, it can be written in terms of two real functions, $\Psi_R(\vec{r}, t)$ and $\Psi_I(\vec{r}, t)$,

$$\Psi(\vec{r}, t) = \Psi_R(\vec{r}, t) + i\Psi_I(\vec{r}, t), \tag{3.4}$$

which we call the 'real' and the 'imaginary parts' of Ψ, respectively. Only the modulus of Ψ, defined by

$$|\Psi(\vec{r}, t)|^2 = (\Psi_R(\vec{r}, t))^2 + (\Psi_I(\vec{r}, t))^2 \tag{3.5}$$

has a physical significance. It gives the probability density of finding the particle at the point $\vec{r}$ at time t. From a mathematical point of view, it is clear that, if a function $\Psi_1(\vec{r}, t)$ is a solution of Schrödinger's equation, so is any other function of the form $\Psi_2(\vec{r}, t) = C\Psi_1(\vec{r}, t)$, with C an arbitrary complex number. We say that Schrödinger's equation determines the wave function only up to a multiplicative constant. This arbitrariness is restricted by the probabilistic interpretation of Ψ. The particle must be somewhere with certainty at time t, therefore the wave function must satisfy the *normalisation condition*

$$\int |\Psi(\vec{r}, t)|^2 d\vec{r} = 1. \tag{3.6}$$

It follows that the modulus of the constant C must be equal to one: $|C|^2 = 1$. It is easy to see that a constant which satisfies this unimodularity condition is given by a phase,

$$C = e^{i\theta} = \cos\theta + i\sin\theta, \tag{3.7}$$

with θ an arbitrary angle $-\pi < \theta \leq \pi$. Consequently, the wave function is determined up to a phase.

Fock searched for the forces necessary to make this invariance *local*. The answer, which is rather easy to obtain, is that they are the electromagnetic forces. Schrödinger's equation with a local phase invariance, i.e. an invariance in which the phase of the wave function θ is not a constant, but an arbitrary function of $(\vec{r}, t)$, describes the motion of a charged particle in an electromagnetic field. Electromagnetism also has a geometric origin, although here the geometry is not that of ordinary space.

After this first success, we would expect that physicists would have followed the same method to construct the gauge invariant theory corresponding to isotopic spin symmetry, immediately after the latter was introduced by Heisenberg in 1932. But here history took an unexpected turn. The fascination that general relativity had exerted on that generation of physicists was such that, for many years, physicists were unable to conceive local transformations of an internal symmetry space without, at the same time, also introducing those of ordinary space.[12] It was only in 1954 that Chen Ning Yang and Robert Mills (see Figure 3.8) wrote the theory known under their name and which is the gauge theory invariant under local isotopic spin transformations. But then we had to face a new problem, that of mass.

[12] Attempts in this direction were made by the Swede, Oscar Benjamin Klein (1894–1977) in 1937 and the Austrian Wolfgang Ernst Pauli (see Figure A1.7) in 1953.

4

A Problem of Mass

The concept of mass in elementary particle physics, introduced in Appendix 1, is not different from the one we know in classical physics. We have the *inertial mass*, which is the parameter entering Newton's equations, and the *gravitational mass*, which determines the coupling of the particle with the gravitational field. The principle of equivalence tells us that the two are equal. All known particles have non-zero masses, with the exception of the photon and, we believe, the graviton. But, as we have said in Chapter 2, we have good reasons to believe that this has not always been the case in the early Universe. There was a time in which most of the elementary particles had zero mass. In this chapter we want to explain the reasons.

4.1 Mass and the range of interactions

One of the basic principles of quantum theory is the duality between a field and a particle. It goes back to de Broglie's founding work in which he postulated that for every particle there is a corresponding wave. This was further enriched by the interpretation of Niels Bohr and the Copenhagen School, according to which these two aspects are dual to each other: for every phenomenon there are experiments which illustrate the corpuscular aspects and others which do the same for the wave aspects.

Let us take the example of electromagnetic forces. In classical physics, the interaction between two charged particles at a distance R apart is described by the electromagnetic field created by either of them at the position of the other. In quantum physics, to this picture we superpose 'a dual one', according to which the interaction is the result of the exchange of one, or more, *photons*, the particle that corresponds to the electromagnetic field. Figure 4.2(a) shows a graphic representation for the case of the interaction of an electron and a proton: one of the two emits a photon which is absorbed by the other. We say that the photon is the 'mediator' of electromagnetic interactions.

In 1935 Hideki Yukawa (Figure 4.1) extended this concept of the forces generated by particle exchange, to nuclear forces. He postulated the existence of a new particle, *the π meson,* or *pion,* which would be the mediator of the nuclear forces. Figure 4.2(b) shows the one pion exchange process between a proton and a neutron. The experimental discovery of the pion in cosmic rays, in 1947, confirmed this idea which became one of the cornerstones in the physics of elementary particles. As we explain in Appendix 1, at present we have identified experimentally the mediators of three out of the four fundamental interactions among elementary particles, to wit the strong, the electromagnetic, and the weak

Figure 4.1 Hideki Yukawa (1907–81, Nobel Prize 1949).

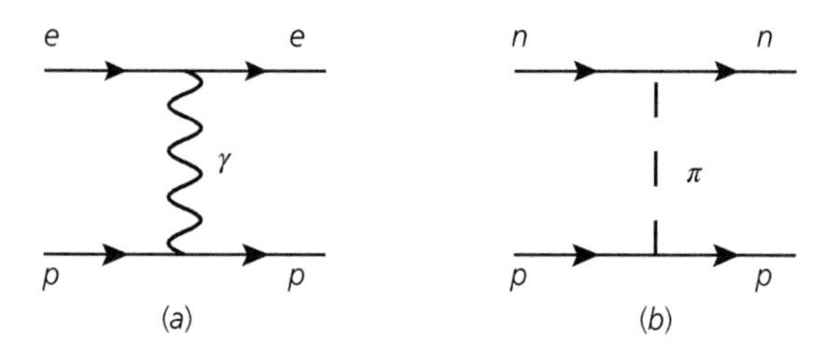

Figure 4.2 Forces produced by particle exchange: (a) electromagnetic forces between two charged particles; (b) the nuclear forces between a proton and a neutron.

interactions. The existence of a mediator for the gravitational interactions, the 'graviton', remains as conjecture.

What is the meaning of such exchanges? Under which form are these particles exchanged? Let us look at Figure 4.2(b). A proton emits a pion which is absorbed by a neutron. The emission process is $p \to p + \pi$. In the initial proton's rest frame the energy of the state equals m_p, the proton mass.[1] The energy of the final state is larger than $m_p + m_\pi$, the sum of the masses of the two particles. It follows that such a process is forbidden by energy conservation. We often say that it is a *virtual process*, although the term is not very meaningful. It becomes a bit more precise in quantum mechanics because of a relation we can write as $\Delta E \Delta t \geq 1/2$, which means that the energy of a state is defined up to an uncertainty ΔE which becomes larger when the time of measurement becomes shorter.[2] We say that for a sufficiently short time the quantum

[1] We recall that we are using a system of units in which the speed of light c and Planck's constant h devided by 2π are both equal to one $c = \hbar = h/(2\pi) = 1$.

[2] We often call this relation the 'time–energy uncertainty relation', but this is very misleading. This terminology is reminiscent of the position–momentum uncertainty relations $\Delta x \Delta p \geq \hbar/2$, which are fundamental in quantum mechanics. There is nothing comparable for ΔE and Δt. Time is not an observable associated with a particle. The meaning of the time–energy relation depends on the kind of measurement we have in mind. For energy we often make frequency measurements of the associated wave packet and, in classical wave mechanics, we have the relation $\Delta \nu \Delta t \geq 1/2$. Quantum mechanics enters via the fact that

fluctuations of the energy become important and the emission process becomes possible.

These remarks indicate that there exists a relation between the range of an interaction and the mass of the exchanged particle. For a large mass, ΔE is large and therefore Δt must be short. A short time implies a short range for the interaction. Although the above arguments were hand-waving, it is straightforward to derive a rigorous and precise relation, first obtained by Yukawa: the exchange of a particle with mass m produces a potential, *the Yukawa potential*, given by

$$V(r) = \frac{e^{-mr}}{r}, \tag{4.1}$$

where V is the potential and r the distance. We see that V decreases exponentially fast with distance with a characteristic constant equal to $1/m$. We also notice that, for $m = 0$, this constant goes to infinity and we recover the Coulomb potential $1/r$. The long range of electromagnetic interactions is related to the zero mass of their mediator, the photon.[3]

4.2 Gauge interactions

In Chapter 3 we introduced the notion of a transformation in the form of a change of the coordinate system. This refers to either the space-time coordinates, or to those of an internal space. If the laws of physics remain invariant under such a transformation, we say that we have a *symmetry*. We also introduced

the energy is proportional to frequency using Planck's constant. It would have been more correct to write this relation as $\Delta E\, T \geq \hbar/2$, where T represents the time interval during which the energy of the system varies significantly. For example, for an atom in its ground state, T can be as large as we wish and the energy can be determined with arbitrary precision.

[3] This result, as well as many others we shall derive in this book, is specific to a three-dimensional space. For example, the electrostatic field created by two opposite charges $+q$ and $-q$, in a one-dimensional space remains constant, independent of the distance.

the notion of a *global* transformation, which is the same for all points of space-time, as well as that of a *local*, or *gauge transformation*. A gauge transformation is a generalisation of a global one in which the parameters of the transformation depend on the point of space-time. Figure 3.9 shows the case of space translations $\vec{r} \rightarrow \vec{r} + \vec{\alpha}(\vec{r}, t)$. The parameter is the vector $\vec{\alpha}$ which determines the translation and, for a gauge transformation, it is an arbitrary function of the space-time point $(\vec{r}, t)$. We explained that invariance under local transformations implies the presence of forces. We indicated, without proving it, that the forces corresponding to local translations are those of gravitation. Here we want to pursue this idea a bit further; staying with Figure 3.9, let us consider two points in space P_1 and P_2, at a distance R apart. We can imagine a function $\vec{\alpha}(\vec{r}, t)$ which vanishes everywhere except at the vicinity of each one of the two points. The translation of the straight line A of Figure 3.9 will give the curve $\tilde{A}$ of Figure 4.3. We see that the gravitational forces must have a range at least as large as the distance R between the points P_1 and P_2. Since this distance can be chosen as large as we wish, we conclude that the invariance under local translations implies a long range for the gravitational forces. By the previous argument this means that their assumed mediator, the graviton, must have a zero mass.

It is easy to generalise this argument to all forces produced by gauge symmetries. They are all produced by the exchange of mediators which in Appendix 1 we called *gauge bosons*. Since by the previous argument the forces must have a long range, the gauge bosons must be massless. In fact we have as many such

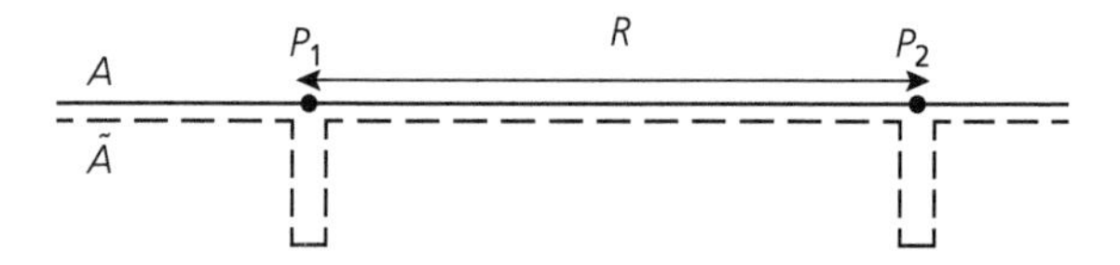

Figure 4.3 Under a space translation with a vector $\vec{\alpha}$ which is different from zero only at the vicinity of the points P_1 and P_2, the image of a free particle trajectory A is the curve $\tilde{A}$.

massless particles as independent gauge transformations which are symmetries of our system. It is a direct consequence of the geometric properties of gauge theories which establish correlations between points at arbitrarily large distances. This result is in perfect agreement with experiment for the two interactions of classical physics, gravitation and electromagnetism, which have, indeed, a long range. Contrary to this, it is in violent disagreement with the other two interactions of microscopic physics, to wit the strong and the weak interactions, whose ranges are measured to be 10^{-15} and 10^{-18} m, respectively. This is the first, and the most fundamental *problem of mass* which has haunted gauge theories since their formulation by Yang and Mills in 1954 up to their first successful application by Steven Weinberg in 1967.

4.3 The masses of matter constituents: quarks and leptons

In Appendix 1 we introduced two kinds of elementary particles: force mediators and matter constituents. We also indicated that particles with integer spin are *bosons* and those with half-integer spin are *fermions*. The force mediators are bosons and we have just seen that gauge invariance seems to force them to be massless. We want to show here that the same result applies, albeit for different reasons, to the matter constituents, quarks and leptons, which are fermions.

4.3.1 Chirality

The constituents of matter are all particles with spin equal to 1/2 (see Appendix 1). We have already noticed that spin is a vector. In classical mechanics the projection of a vector on an axis equals the vector's modulus times the cosine of the angle between the vector and the axis. A remarkable result of quantum mechanics is that the projection of a spin can take only discrete values: for the

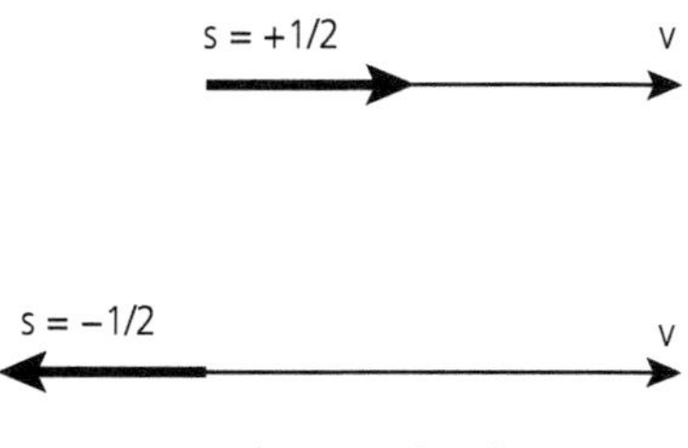

Figure 4.4 The two chirality states.

electron they can be $+1/2$ or $-1/2$, independent of which axis we choose.[4]

For a spin 1/2 particle, such as an electron or a quark, moving with a speed $\vec{v}$, it is instructive to choose as a projection axis the direction of $\vec{v}$. We call the resulting projection *chirality*. In Figure 4.4 we present the two possible values: $+1/2$ when the spin projection is parallel to $\vec{v}$ and $-1/2$ when it is opposite.

In the physics jargon these two states are called *right chirality* and *left chirality*,[5] respectively. Chirality is preserved by rotations, because both vectors, spin and speed, turn the same way. Conversely, the two states are exchanged by space inversion, or parity, which we introduced in Section 3.1.3. We can understand this by recalling that spin has the properties of angular momentum, which in classical mechanics is proportional to the vector product of the position and the speed vectors $\vec{L} \sim \vec{r} \times \vec{v}$. Under parity both change sign, so the angular momentum does not. Since speed changes sign, we conclude that parity changes an electron with right chirality to one with left.

These two states can also be connected with a second set of transformations. Consider an electron with speed $\vec{v}$ and chirality $+1/2$. We can perform a Lorentz transformation and go to the rest frame of the electron, i.e. the frame in which its speed vanishes.

[4] For a spin s, integer or half-integer, the possible values of the projection are $s, s-1, s-2, \ldots, -s$.

[5] The terminology comes from the two states of circular polarisation of light, although the term 'chirality' is used only for fermions.

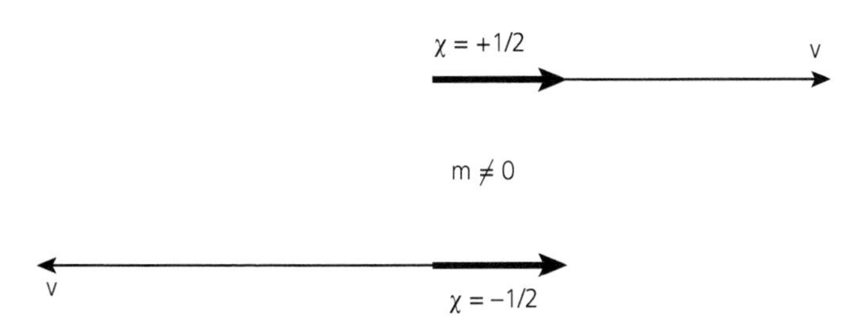

Figure 4.5 For a massive particle the two chirality states are connected by Lorentz transformations.

The spin projection remains unchanged. With a second Lorentz transformation we can go to the system in which the speed of the electron equals $-\vec{v}$ (see Figure 4.5). The combination of these two transformations allows us to change a particle with right chirality to one with left chirality, under one condition: the mass of the particle must be non-zero. Indeed, a massless particle always travels with the speed of light and there exists no transformation which can bring it to rest.

4.3.2 *Chirality and weak interactions*

In this section we will talk only about weak interactions. We recall that they are responsible, in particular, for β-decay of the neutron of the form

$$n \rightarrow p + e + \bar{\nu}_{(e)}, \tag{4.2}$$

where n denotes the neutron, p the proton, e the electron, and $\bar{\nu}_{(e)}$ the anti-particle of the neutrino associated with the electron.[6] We find the same reaction in nuclear physics where the neutron and the proton are bound inside the nuclei. In the inverse direction, this gives rise to the fusion of two protons (hydrogen

[6] In Appendix 1 we indicate that we have three species of neutrinos, each being associated with one of the three known leptons, the electron (e), the muon (μ), and the tau (τ).

nuclei), giving a nucleus of deuterium, D, which is a bound state of a proton and a neutron,

$$p + p \rightarrow D(p + n) + e^{+} + \nu_{(e)}, \tag{4.3}$$

where e^{+} is the positron, the anti-particle of the electron. This reaction is the starting point of the process of nucleosynthesis in the interior of stars. They are reactions which generate all heavy elements and are at the origin of stellar energy. We can rewrite these reactions in terms of quarks by replacing the proton and neutron by quarks u and d respectively. We can also generalise them with the quarks of other families and describe the decays of several unstable particles.

In this section we want to present some properties of weak interactions which distinguish them in a very significant way from all other interactions.

Among all interactions of elementary particles, the only ones that do not respect the invariance under space inversions, are the weak interactions. This is an experimental fact, not a theoretical result.

A second property of these interactions, also established by experiment, is that they involve primarily fermions, quarks or leptons, of left chirality. The states with right chirality seem to be insensitive to weak interactions. We shall come back to this point in Section 6.4.

These two properties make weak interactions unique. We see that, in order to describe them correctly we must assume (i) the violation of parity invariance and (ii) a zero mass for all constituents of matter, leptons and quarks. If (i) is rather easy to implement, (ii) is in contradiction with all experimental results which show that all these particles have non-zero mass. It is the second *problem of mass* that we encounter. It appears to be less 'fundamental' than the first, which refers to the gauge bosons which transmit interactions. The first was a consequence of the geometric property of gauge interactions, while the second seems to have a phenomenological origin. Nevertheless, they are both very important and they call for profound modification of our ideas on symmetries and the four interactions we introduce in Appendix 1. It is this part of the story that we shall present in the following chapters.

5

Spontaneously Broken Symmetries

5.1 Curie's theorem

In this chapter we shall present a phenomenon which seems to contradict our physical intuition, although, as we shall see, we encounter it in everyday life. It is the phenomenon of *spontaneous symmetry breaking*.

Usually, for a symmetric problem, we are looking for symmetric solutions. We cite Pierre Curie:

> 'When certain causes produce certain effects, the symmetry elements of the causes must be present in the produced effects'

Talking about spontaneous symmetry breaking seems to contradict this assertion. Of course, Curie knew about these phenomena.

Let us take an example. Consider a sphere of radius r uniformly charged with an electric charge Q. We want to compute the electrostatic field produced at a point P at a distance $R > r$ from the centre of the sphere (see Figure 5.1).

In order to solve this problem it is sufficient to notice that 'the cause', i.e. the charged sphere, has a spherical symmetry. Consequently, the same will be true for 'the effect', in other words, the field $\vec{E}$ will be radial because it is the only direction which respects the spherical symmetry. Furthermore, for the same symmetry reasons, the field must have the same absolute value at any point on the spherical surface going through P with centre at the centre of the sphere. It is now straightforward to

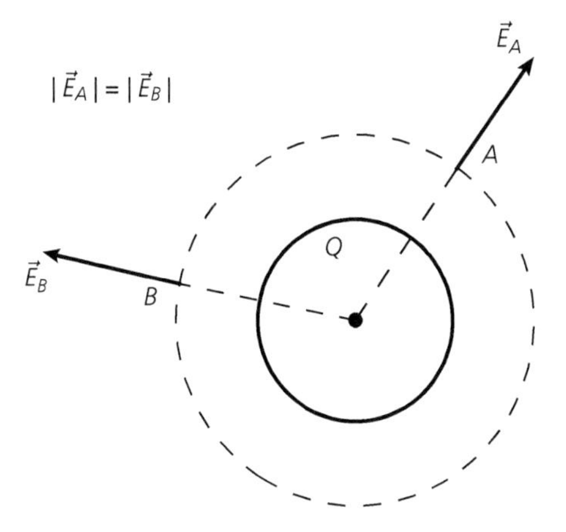

Figure 5.1 The electric field produced by a uniformly charged sphere at a distance R.

compute $\vec{E}$ by applying a theorem by Carl Friedrich Gauss which gives: $\vec{E} = (Q/4\pi\epsilon_0 R^2)\vec{r}$, where $\vec{r}$ is the unit vector in the radial direction and ϵ_0 is the dielectric constant.[1]

In practice a sphere is never perfect and the symmetry could only be approximate. Nevertheless, we still apply the above reasoning assuming implicitly that 'small' violations of the symmetry of the cause will induce only 'small' deviations from the symmetric solution. But this is a much stronger assumption going beyond Curie's theorem. Indeed, the latter shows only *the existence* of a symmetric solution, while in practice we may need in addition an assumption on its *stability*. In the case of spontaneous symmetry breaking we shall study here, it is this last assumption which will fail.

[1] This is a special case of the general Gauss theorem, which we shall not prove here. Consider an arbitrary distribution of static electric charges in a finite region of space and a closed surface which surrounds them. The theorem states that *the flux* of the electric field through the surface, i.e. the integral over the surface of the perpendicular component of the field, equals the sum of the charges divided by ϵ_0.

5.2 Spontaneous symmetry breaking in classical physics

A simple example is shown in Figure 5.2. A cylindrical rod of length l is loaded with a load F along the z axis. The problem is symmetric under rotations around this axis, so we expect the final state to be a compressed but straight rod. This is in fact the answer when F is small. But we know from experience that when F increases, the form of the rod changes: it bends in an arbitrary direction. This is the phenomenon of *buckling*. The bent rod no longer has the rotational symmetry because the bending has introduced a spacial direction in the (x, y) plane. This class of phenomena we call *spontaneous symmetry breaking*.

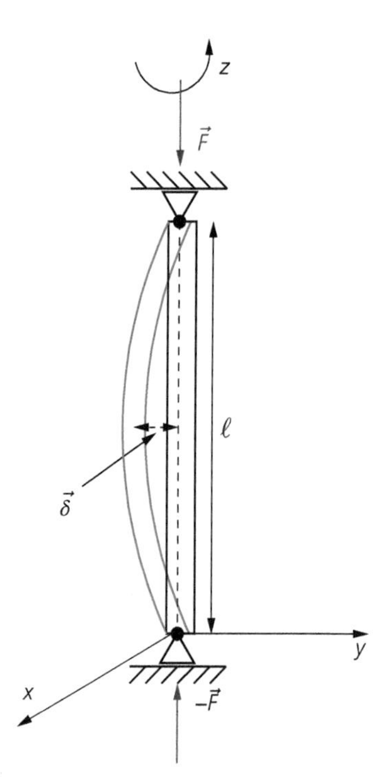

Figure 5.2 The rod bends under the action of the load F.

Question: what happened to the original symmetry? Answer: it is still there, but it is hidden. We cannot predict the direction of the bending. They are all equally probable; in other words, we have an infinite number of possible solutions, all connected via the transformations of the original symmetry, i.e. the rotations around the z axis.

The mathematical analysis of this example is quite simple and we present it in Box 5.1. This is required for quantitative computations, but it is not necessary for a qualitative understanding of the phenomenon. We can easily guess the essential features:

1. For small values of the load the rod stays straight but it starts being compressed. This compression costs a certain elastic energy which increases with the compression rate, therefore, with F.
2. There exists a critical value of the load, F_{cr}, above which it costs less in energy to the rod to bend and decrease its compression rate (a bent rod has a greater effective length and, consequently, a smaller compression rate). The analytical computation we present in Box 5.1 gives the value of F_{cr} precisely, in terms of the rod's parameters for the configuration in Figure 5.2.
3. Theoretically, there always exists the possibility of having a straight but very compressed rod, but, since the associated elastic energy is very large, this is an unstable state.
4. It is convenient to introduce a parameter $\vec{\delta}$. This is a vector in the horizontal (x, y) plane and its components δ_x and δ_y give the displacement of the centre of the rod from the symmetric point. We can compute the elastic energy as a function of δ_x and δ_y, and we obtain Figure 5.3. For $F < F_{cr}$, we obtain the system on the left with a single minimum energy configuration. It is the symmetric solution with $\vec{\delta} = 0$. For $F > F_{cr}$, we obtain the system on the right. The symmetric solution is always present, which means that

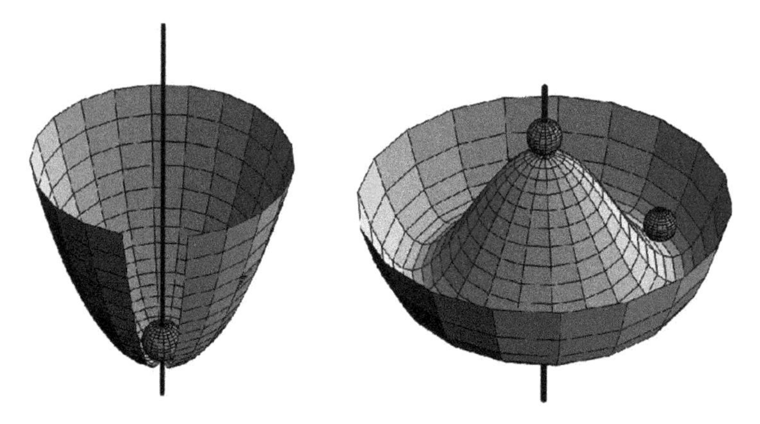

Figure 5.3 The energy of the system as a function of the order parameter, for $F < F_c$ (left) and $F > F_c$ (right). Dr A. Hoecker, CERN.

Curie's theorem is verified, but it corresponds to a local maximum of the energy and it is, therefore, unstable.

5. We have an infinity of stable solutions. The energy depends only on the modulus of $\vec{\delta}$. Starting from any one of these solutions we obtain any one of the others by applying a rotation around the z axis. In the physics jargon, a state of a system with the minimum energy is called *a ground state.* Here we have an infinite number of ground states. We say that the ground state *is infinitely degenerate.*

Box 5.1: The phenomenon of buckling

We present here a simplified version of the mathematical analysis of buckling without entering into all the details of the theory of elasticity. The reader who is not interested in these technical points can go directly to Section 5.3.

Let us call $X(z)$ and $Y(z)$ the x and y displacements of the point z on the axis of the rod from its symmetric position. z varies from 0 to l. In terms of these variables the order parameter is given by

$$\delta_x = X(z = l/2); \quad \delta_y = Y(z = l/2). \tag{5.1}$$

In the symmetric phase we have $X(z) = Y(z) = 0$. The general equations of elasticity are non-linear differential equations which we can only solve numerically. However, we can simplify them in the neighbourhood of the critical point, keeping only first-order terms in X and Y. For the rod of Figure 5.2, which is articulated at both ends, they take the form,

$$IM\frac{d^2X}{dz^2} + FX = 0; \quad IM\frac{d^2Y}{dz^2} + FY = 0, \tag{5.2}$$

where $I = \pi R^4/4$ is the moment of inertia of the rod, R is the radius of its section, and M a parameter which characterises its elastic properties and is called the *Young modulus.* This system of equations is invariant under rotations around the z axis, with an angle θ under which the variables X and Y transform as

$$X \to X\cos\theta + Y\sin\theta; \quad Y \to -X\sin\theta + Y\cos\theta. \tag{5.3}$$

We want to solve system (5.2) with the following boundary conditions at $z = 0$ and $z = l$: $X = Y = 0$, which impose on the extremities to remain fixed, and $X'' = Y'' = 0$ which correspond to an articulated rod. It is clear that the symmetric solution $X = Y = 0$ is always present. However, we have also asymmetric solutions of the form $X = C\sin kz$ with $k^2 = F/MI$, provided $kl = n\pi$; $n = 1, 2, \ldots$ The first of these solutions with $n = 1$ appears as soon as F takes the critical value

$$F_{cr} = \frac{\pi^2 MI}{l^2}. \tag{5.4}$$

Starting from this solution we can obtain an infinity of others by applying the rotation (5.3).

The appearance of these asymmetric solutions at the vicinity (i.e. at first order in the displacements X and Y) of the symmetric solution is a sign of the latter's instability. We can verify this by an explicit computation of the corresponding elastic energy, which we shall not present here. However, it will be instructive to guess its form as a function of the order parameter.

Rotational invariance imposes a dependance on $\vec{\delta} \cdot \vec{\delta} = \rho^2$. At the vicinity of the critical point δ is small and we can write an expansion of the form $E = C_0 + C_1\vec{\delta} \cdot \vec{\delta} + C_2(\vec{\delta} \cdot \vec{\delta})^2 + \ldots = C_0 + C_1\rho^2 + C_2\rho^4 + \ldots$, neglecting higher-order terms. The Cs are constants depending on the parameters of the rod and the load F. The ground state of the system is determined by the minimum of the energy, which gives, in terms of the variables defined in (5.8),

$$\frac{dE}{d\rho}(\rho = v) = 0 \Rightarrow v(C_1 + 2C_2v^2) = 0 \qquad (5.5)$$

This equation has two solutions: the symmetric one, $v = 0$ and a second one, $v^2 = -C_1/2C_2$. ρ being a real number, this second solution is acceptable only if the ratio C_1/C_2 is negative. It is the solution with spontaneous symmetry breaking. If C_2 is negative there is no minimum energy solution; for large ρ, $E \to -\infty$. It follows that the existence of a ground state imposes $C_2 > 0$, in which case C_1 must vanish at the vicinity of the critical point and we can write it as $C_1 = \hat{C}_1(F_{cr} - F)$, with $\hat{C}_1$ positive. The phase with spontaneous symmetry breaking is obtained for $F > F_{cr}$ and, in this phase, we can write the energy as

$$\begin{aligned} E &= C_0 + \hat{C}_1(F_{cr} - F)\vec{\delta} \cdot \vec{\delta} + C_2(\vec{\delta} \cdot \vec{\delta})^2 \\ &= \hat{C}_1(F - F_{cr})\frac{(\rho^2 - v^2)^2}{2v^2}, \end{aligned} \qquad (5.6)$$

where v is given by the non-zero solution of equation (5.5). The energy being defined up to an arbitrary additive constant, we have determined C_0 by the condition of having the ground state ($\rho = v$) energy vanishing. In the phase with spontaneous symmetry breaking, the energy of the symmetric solution $\rho = 0$ is positive and given by

$$E_0 = \hat{C}_1(F - F_{cr})\frac{v^2}{2}. \tag{5.7}$$

This expression gives the form of Figure 5.3. For $F < F_{cr}$ we have a single minimum with $\delta = 0$ (Figure 5.3 left), and for $F > F_{cr}$ we obtain the circle of Figure 5.3 right.

We just described an example of a class of phenomena known as *phase transitions.* The stressed rod is a physical system which may exist in two different phases: the phase of the straight rod and that of the bent rod. We shall call the first one the *symmetric phase* and the second the *broken symmetry phase.* There exists a parameter, external to the system, which determines which one of the two phases will be chosen by the system. In the case of the rod it is the load F. We call this the *control quantity.* The two phases differ by the value of another parameter, for the rod the vector $\vec{\delta}$, which we call the *order parameter.* Its value is zero in the first phase and different from zero in the second. The transition from one phase to the other is accompanied by a change of the symmetry of the system. In the case of the rod it is the rotational symmetry around the z axis which seems to be absent in the second phase. As we have pointed out, the symmetry is hidden because the ground state is degenerate.

The order parameter will play an important role in our discussion of spontaneous symmetry breaking. It will be convenient to parametrise $\vec{\delta}$ by its modulus ρ and a phase θ, rather than its components δ_x and δ_y. We introduce a complex number $\delta = \delta_x + i\delta_y$ with

$$\delta_x = \rho\, cos\theta; \quad \delta_y = \rho\, sin\theta; \quad \delta = \rho\, e^{i\theta}. \tag{5.8}$$

With these variables the points on the minimum energy circle correspond to ρ constant and arbitrary θ, between $-\pi$ and π.

We have many examples of spontaneous symmetry breaking in classical physics, such as crystallisation, turbulence, many problems of erosion, etc.

5.3 Spontaneous symmetry breaking in quantum physics

Ferromagnetism, i.e. iron magnetic properties, offers an example of spontaneous symmetry breaking in quantum physics. The phenomenon is as follows: looking at the magnetic properties of an iron rich metal we often find two phases. We shall describe them using the terminology we introduced in the case of the bent rod.

1. The symmetry is that of three-dimensional rotations.
2. The *control quantity* is the temperature T. There exists a critical value T_c, called the *Curie point*. For $T > T_c$ we are in the symmetric phase, and for $T < T_c$ in that with spontaneous symmetry breaking. For iron the Curie point is on the order of 770 °C.
3. The *order parameter* is the *magnetisation*. It is a macroscopic magnetic moment which determines the interaction of the sample with a magnetic field. The simplest compass is precisely a needle made out of magnetised iron which points parallel to the earth's magnetic field. Above the Curie temperature the magnetisation vanishes. At $T < T_c$ a magnetisation appears spontaneously.
4. In the high temperature phase there is no magnetisation (*para-magnetic phase*). There is no privileged direction in space and we have the full three-dimensional rotation symmetry. This is the symmetric phase. In the low temperature phase we have a spontaneous magnetisation (*ferro-magnetic phase*). It defines a privileged direction in space. The symmetry is reduced to rotations only around this axis. This is the broken symmetry phase.

5. In the ferro-magnetic phase the ground state of the system is degenerate. All orientations of the magnetisation are *a priori* equivalent.

W. Heisenberg proposed a simple model which, in a series of approximations, eliminates all inessential complications and captures the main features of the phenomenon.

The first approximation concerns the crystal structure. We assume a regular lattice and we ignore all possible defects. In order to simplify the presentation we take a cubic lattice, although the reasoning also applies to other regular crystals.

The iron atoms occupy the lattice sites. In principle we must take into account all atomic degrees of freedom, but, and this is the second approximation, we assume that, for the magnetic properties of the crystal, only the spin degrees of freedom are important. We consider a spin in every lattice site, for example a spin equal to 1/2, whose projection on the z axis can take the two values $\pm 1/2$.

The spins interact among themselves and, following Heisenberg, we assume the interactions to be of short range so that only pairs of nearest neighbours are coupled. For a three-dimensional cubic lattice, each spin has six nearest neighbours.

The last assumption concerns the form of the interaction. We assume rotational invariance and, the simplest form of coupling yields an energy given by

$$E = -\frac{1}{2}J\sum_{i,j}\vec{S}_i \cdot \vec{S}_j, \tag{5.9}$$

where $\vec{S}_i$ is the spin on the i_{th} site and the sum runs over all pairs of nearest neighbours. J is a constant, which we assume to be positive.[2] Rotational invariance is ensured by the form of the interaction which is a scalar product of two vectors.

[2] The case with $J < 0$ is also interesting and describes a phenomenon called *anti-ferromagnetism*. It has been studied by Louis Eugène Félix Néel (1904–2000, Nobel Prize 1970), but we shall not present it here.

We can solve this model numerically and in some simple cases, for one- or two-dimensional crystals, analytically, but the physics of the phenomenon is obvious by inspection.

At very high temperature we are in the symmetric phase. Thermal fluctuations are very important and, at any particular moment, the spins are randomly oriented. We obtain the configuration of Figure 5.4 (a) with vanishing mean value of the magnetisation. This is a disordered phase.

At very low temperature, when thermal fluctuations can be neglected, the ground state of the system is obtained by the configuration which minimises the energy. For positive J this corresponds to maximising the scalar product, which means making the spins parallel. Each spin tends to orient its nearest neighbours in its own direction, so that, from nearest neighbour to nearest neighbour, we obtain the configuration of Figure 5.4 (b), yielding a non-zero magnetisation $\vec{M}$. This is the ordered phase in which the original three-dimensional rotation symmetry is spontaneously broken.[3]

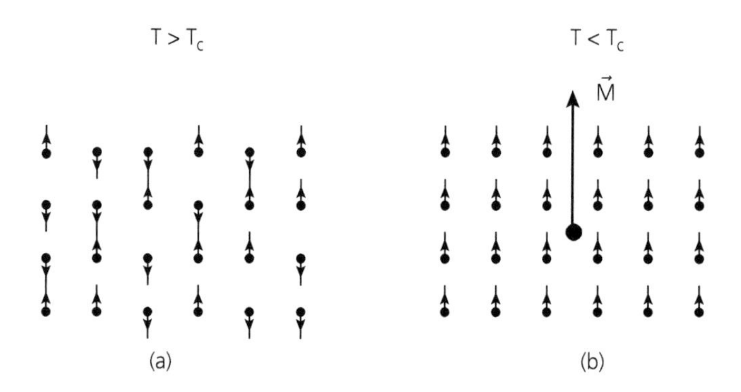

Figure 5.4 (a) At $T > T_c$ the spins are oriented randomly due to the thermal fluctuations. The average magnetisation vanishes. (b) : At $T < T_c$ the interaction orients the spins and we have a spontaneous magnetisation.

[3] What happens in a real ferromagnet is more complex. This process of spin orientations starts independently from various points of the solid and the image shown in Figure 5.4 (b) is valid only locally in a small domain. A sample of

To summarise, at every temperature, we have two competing effects: the thermal fluctuations favour disorder but their strength decreases with temperature; the interaction favours order. We can understand that there exists a temperature below which order wins. This is the Curie temperature.[4]

A remark before closing this section: the order parameter of ferromagnetism is the vector of magnetisation $\vec{M}$. In the high temperature phase $\vec{M} = 0$ and the system is invariant under the full set of three-dimensional rotations. This is the symmetric phase. An arbitrary rotation can be parametrised by the three Euler angles, i.e. two angles to define the direction of the axis around which we perform the rotation, and a third which gives the rotation angle. In the low temperature phase the three-dimensional rotational symmetry is not completely broken. A part, corresponding to the rotations around the $\vec{M}$ axis, remains exact. Consequently, we need two angles to label the states of minimum energy. In the notation we used in the case of the bent rod, we write

$$M_x = \rho \sin\theta \cos\phi; \quad M_y = \rho \sin\theta \sin\phi; \quad M_z = \rho \cos\theta. \quad (5.10)$$

We see that in the low temperature phase, (i) the symmetry is only partly broken and (ii) the set of ground states form the surface of a sphere. In Chapter 6 we shall encounter cases in which

macroscopic size contains a large number of domains, each with a magnetisation pointing in a random direction. This is the reason why a piece of iron does not always appear to be magnetised. Strictly speaking, this domain structure does not minimise the energy because in the boundaries between domains the spins are not parallel. Nevertheless, transition to the real ground state would require a simultaneous change of orientation of the spins in every domain and the probability of such an event goes to zero for a large sample. In order to obtain a magnetised piece of iron we must place the sample in an external magnetic field during the phase transition.

[4] Another result which depends on the number of space dimensions. We can find a qualitative argument which explains this dependence: the interaction, which favours order, increases with the number of nearest neighbour pairs, i.e. the number of space dimensions. For example, we can show that a linear chain of spins, in which every spin has only two nearest neighbours, has no phase transition for any finite value of the temperature.

this set forms a multi-dimensional hyper-surface. Notice also that in the low temperature phase, which is the ordered phase, the energy of the symmetric solution is given by the analogue of formula (5.7), but multiplied by a factor proportional to the total number of spins. This goes to infinity for an infinite system.

5.4 Goldstone theorem

In this section we want to present a very important consequence of spontaneous symmetry breaking which, in addition, will bring us closer to the subject of this book, namely the Brout-Englert-Higgs particle. This is a quantum effect but, as we have done up to now, we shall try to find classical analogies, with all the dangers of offering misleading explanations inherent in such an approach.

Let us look at Figure 5.4 (b). At low temperature the system is in the ordered phase, all the spins are parallel, and we have long range correlations. In Chapter 4 we showed that in quantum physics and in a three-dimensional space, such correlations imply long range interactions and, consequently, massless particles. This is *Goldstone's theorem:*[5] *spontaneous symmetry breaking*[6] *results in the appearance of massless particles in the spectrum of the theory.*

We shall not give here a mathematical proof of this theorem; we shall try instead to extract its physical consequences.

The first image is the degeneracy of the ground state of a system in the phase with spontaneous symmetry breaking. It is represented classically by the right-hand side of Figure 5.3. All points in the circle $\rho = |\delta| = v$, where v is the radius of the circle, correspond to the same value of the energy. We want to find the corresponding representation in quantum physics. Two simple principles will help us:

1. For every configuration of classical variables there is a corresponding state in quantum physics. This is a direct consequence of the axioms of quantum mechanics.

[5] It is also called the Nambu–Goldstone theorem in honour of Yoichiro Nambu and Jeffrey Goldstone (see Figure 5.5).

[6] The theorem applies only to continuous symmetries.

Figure 5.5 Spontaneous symmetry breaking: Pierre Curie (1859–1906, Nobel Prize 1903); Yoichiro Nambu (1921–, Nobel Prize 2008); Jeffrey Goldstone (1933–). Goldstone : J. Goldstone archives.

2. In quantum physics of elementary particles, each state is characterised by the number and the properties of the particles it contains.

With the help of these two principles we want to find the classical analogue of Goldstone's theorem. We have already alluded to the field–particle duality in quantum physics. Therefore we expect to find the fields which correspond to the Goldstone particles and, in the classical limit, to identify the corresponding classical variables.

Let us start with Figure 5.3. There are two relevant variables, ρ and θ. A classical configuration is described by precise values of these two variables. According to the aforementioned first principle, we also have a quantum state. On the other hand, going from classical to quantum physics, classical variables become quantum fields which correspond to particles. We conclude that the quantum description of spontaneous symmetry breaking will involve two kinds of particles, those which correspond to ρ and those to θ. According to the second principle, each state contains a different number of these particles.

Consider two quantum states, each corresponding to a different point in the circle of minima in Figure 5.3. In classical physics they differ only by the value of θ, therefore, each one of the two quantum states has the same number of ρ particles, but a different number of θ particles. Let us call this difference Δ.[7] The two states must have the same energy. It follows that we can add a number Δ of θ particles in a state without changing its energy. In a relativistic theory the energy of a particle is always larger, or equal, to that given by its mass. Therefore, the mass of the θ particles must be equal to zero. These are the Goldstone particles. We can now state the classical version of Goldstone's theorem:

> *Spontaneous breaking of a continuous symmetry implies the degeneracy of the minimum energy state (ground state) of a physical system.*

This degeneracy is translated, in the quantum system, into the appearance of zero mass particles.

In fact, our analysis also yields a second conclusion. We have seen that the quantum description of spontaneous symmetry breaking involves two kinds of particles, the θs, which are the massless Goldstone particles, and the ρs which are massive.[8] In the symmetric phase the ground state is the one which corresponds to $\rho = 0$. In the phase with spontaneous symmetry breaking we have an infinite number of ground states, all corresponding to $\rho = v$. They differ from the symmetric state by the number of ρ particles. In our jargon we often call the ground state *vacuum*. Therefore, the phase transition implies a change in the vacuum state of the theory. In Section 5.3 we pointed out that, for a large system, the energy difference between these two states goes to infinity. The two 'vacua', the symmetric one and the one with spontaneous symmetry breaking, not only are they not 'empty', but, in addition, they differ by an infinite number

[7] From the mathematical point of view, this is not a rigorous argument. It turns out that each one of the quantum states which corresponds to these classical configurations, contains an infinite number of particles. As a result Δ is not well defined.

[8] We recall that we may have several kinds of 'θ' particles. In the example of ferro-magnetism we had two: the θs and the ϕs.

of ρ particles.[9] In Section 5.5 we shall establish that they are the Brout–Englert–Higgs particles.

A last remark: in Appendix 1 we argue that the mass of a particle may depend on the interactions in which it participates. Consider a particle which interacts with ρ. We expect to find out that its mass may be different in each of the two phases of the system. If we call m_0 its mass in the symmetric phase which has no ρ particles, in the phase with spontaneous symmetry breaking it will be given by

$$m = m_0 + \delta m, \tag{5.11}$$

where δm is the change in the mass that the particle gets from its interactions with ρ.

5.5 Spontaneous symmetry breaking in the presence of gauge interactions

This is the most important section of this book. It is here that the significance of the CERN discovery will be presented.

Let us summarise the results we obtained previously on massless particles. In Chapter 4 we showed that the presence of interactions with local symmetries, i.e. gauge interactions, seems to make the bosons which mediate these interactions massless. This was the first mass problem. Next we discovered a second problem, namely, that the weak interactions also make the constituents of matter, leptons and quarks, massless. In this chapter we have proved Goldstone's theorem which sounds like a third mass problem: the phenomenon of spontaneous symmetry breaking implies the presence of new massless particles, those we called Goldstone particles. However, looking at the table of elementary particles in Appendix 1, we see that all particles with the exception of the photon and, possibly the graviton, are massive. We could

[9] We also see that the bosonic character of the ρ particles is essential. Only bosons can accumulate in large numbers in the same state and it is this property that makes this phenomenon of phase transition possible.

conclude that each of these problems, taken separately, indicates that gauge theories, as well as the phenomenon of spontaneous symmetry breaking, are in contradiction with the experimental results. In this section we will show that, in fact, the opposite is true: if we consider the three problems together, each will solve the other. In a gauge theory with spontaneous symmetry breaking, all particles can be massive. It was the discovery of this property, we could call it 'miraculous', which paved the way for spectacular progress in our understanding of the fundamental interactions.

In this section we shall try to explain this phenomenon. Unfortunately, it is a complex and purely quantum phenomenon, for which we have no classical analogue. As we just said, we have to face three mass problems: (i) the masses of the matter constituents, (ii) those of the gauge bosons, and (iii) the mass of the Goldstone particles, which we previously called 'θ particles'.

The simplest to start with is probably the first one. We have already referred to it at the end of Section 5.4 by writing equation (5.11). We noticed in Chapter 4 that the weak interaction symmetries force quarks and leptons to be massless. As a result, in the symmetric phase we must have $m_0 = 0$. Spontaneous symmetry breaking allows these particles to develop a δm and acquire a mass. This happens through their interactions with the ρ particles which populate the ground state in the broken symmetry phase. We expect δm to be proportional to the strength of the interaction between the particle and ρ and this allows us to fit the large spectrum of masses we observe among the matter constituents. But, beware: *fit* is not synonymous with *understand*. Table 5.1 gives the experimentally determined masses of quarks and leptons. We see that the ratio between the mass of the quark t (the heaviest) and that of the electron (the lightest)[10] is of order 350,000. By

[10] In this book we shall not discuss the problem of the neutrino masses. We know, by indirect measurements, that they are different from zero, but they

Table 5.1 The masses of quarks and leptons. The unit is the megaelectronvolt (MeV). 1 MeV $\sim$ 1.7810^{-30} kg. The values shown are approximate. Quark masses are not directly measurable (see Chapter 6) and those of the neutrinos are too small and beyond our measuring capabilities.

Masses of matter constituents						
Quarks	u	d	c	s	t	b
	2.3 MeV	5.0 MeV	1275 MeV	95 MeV	173,000 MeV	4180 MeV
Leptons	$\nu_{(e)}$	e	$\nu_{(\mu)}$	μ	$\nu_{(\tau)}$	τ
	??	0.51 MeV	??	106 MeV	??	1777 MeV

adjusting the coupling strength of each of these particles with ρ we can reproduce the values we see in Table 5.1, but we do not understand the origin of such a dispersion. This makes us believe that an important part of the mass creation mechanism is still beyond our comprehension.

Let us now come to the second problem, namely the masses of the gauge bosons which transmit weak interactions. In Table A1.3 these are indicated as $W^{\pm}$ and Z^0. Their masses have been measured and found to be 80,385 and 91,188 MeV, respectively. We could be tempted to use the previous argument and attribute their masses in the phase with spontaneous symmetry breaking to their interactions with ρ. This is not totally wrong but it describes only one aspect of the phenomenon. In fact, the real story is more complex.

These gauge bosons have spin equal to one. This follows from a mathematical theorem, which we have not proven here; but it is also an experimental result. In quantum mechanics we show that, for a particle with spin equal to s, the spin projection on any axis can take $2s + 1$ values, to wit $+s$, $+s - 1$, $\ldots$, $-s$. Therefore, for $s = 1$, we have three possible values, $+1$, 0, and -1. We say that a spin 1 particle is described by *three degrees of freedom*. This rule applies

are so small that we have not been able to measure them directly. Present limits indicate that they are at least 500000 times lighter than the electron.

to massive particles. For massless particles the rule is different: for all s different from zero we have only two degrees of freedom and the spin projection on the direction of motion can only take the values $+s$ and $-s$.

These properties result from the relativistic invariance of the theory and we cannot demonstrate them here. We can only give a classical analogue. It is known that electromagnetic waves can be polarised, but only in a transverse direction with respect to their propagation. We never observe them polarised in the longitudinal direction. In the equations of classical electrodynamics, the electromagnetic field has two independent components, corresponding to the two transverse polarisations.[11]

This is the complication we mentioned previously: for the gauge bosons it is not enough to look for interactions to generate mass, we must also find an extra degree of freedom to allow for the transition between a massless and a massive spin one particle. Who supplies this degree of freedom?

In order to answer this question we must come back to the reasons why we have massless particles in the first place, in both gauge theories and the Goldstone phenomenon. We have explained that in both cases this was the result of long range correlations. The gauge interactions impose a long range because the transformations depend, in an arbitrary way, on the space-time point. The spontaneous symmetry breaking imposes a different long range order, as is indicated in the spin example of Figure 5.4(b). It is easy to imagine that the two may not be compatible with each other. For example, it is easy to see in this figure that, if we allow ourselves to turn the spins independently at every point, we can destroy the order imposed by the spontaneous symmetry

[11] There is an exception to this rule: the electromagnetic field propagating in a superconductor has, in addition, a longitudinal component. It is the property which led Philip Anderson to describe, for the first time, the phenomenon we are in the process of studying here, namely that of spontaneous symmetry breaking in the presence of electromagnetic interactions, in a non-relativistic context.

breaking.[12] The absence of long range order implies the absence of massless particles. Therefore we expect that in the phase with spontaneous symmetry breaking, all particles are massive.

Let us see the consequences of this analysis, first on a simple system with a single gauge boson. In the symmetric phase we have the massless gauge boson described by the two degrees of freedom corresponding to its transverse polarisations. In addition, we have two massive spin zero particles, those we called θ and ρ. Altogether four degrees of freedom. In the phase with spontaneous symmetry breaking the same physical system is described by a massive spin one particle (which makes three degrees of freedom corresponding to its transverse *and* longitudinal polarisations), and a massive spin zero particle, the ρ. Again, a total of four degrees of freedom. The θ particle, which would have been massless as a consequence of Goldstone's theorem, is absent. Its degree of freedom has been used in order to allow the gauge boson to acquire a mass. We can say that the θ particle is the component of longitudinal polarisation of the massive spin one boson.

This picture is immediately generalised to a more complex system, like that of ferromagnetic spins in the presence of gauge symmetry. The symmetry is that of three-dimensional rotations. We have three independent transformations and this implies that, in the high temperature phase, we have three massless gauge bosons. This gives us six degrees of freedom. We add the three massive, spinless particles corresponding to ρ, θ, and ϕ. In the low temperature phase we have a partial spontaneous symmetry breaking. The symmetry of rotations around the axis defined by the direction of $\vec{M}$, remains intact. Only the other two are spontaneously broken. It follows that only two among the three gauge bosons acquire a mass; the third remains massless. In this process, the first two gauge bosons absorb the degrees of freedom of the

[12] Again an argument which we should consider as an indication and not as a proof. It is easy to visualise and understand it on a discrete lattice, but its applicability in a continuous space is much more subtle.

particles θ and ϕ. So, in this phase we have: (i) a massless gauge boson with two degrees of freedom, (ii) two massive spin one bosons each having three degrees of freedom, and (iii) the spin zero boson ρ. The sum of degrees of freedom is again nine, the same as in the symmetric phase.

The trace of this phase transition is the presence of the ρ particle. We believe it is the particle discovered at CERN. Therefore this discovery not only adds a new entry to our table of elementary particles, but also, it opens a window into one of the most extraordinary phenomena in the history of the Universe.

6

The Standard Theory

6.1 Introduction

In Appendix 1 we see that all experimental results show the presence of four fundamental interactions among elementary particles: *gravitational interactions, weak interactions, electromagnetic interactions*, and *strong interactions.* In this chapter we want to use the mathematical and conceptual tools we have developed so far in order to build theoretical models describing these interactions. The ingredients of these models are:

1. Gauge symmetries. These are generated by transformations whose parameters depend on the space-time point. In Chapter 3 our motivation for introducing them was essentially aesthetic, but, in fact, they are absolutely necessary. We can prove that the only theories which are mathematically consistent and have the predictive power to correctly describe fundamental interactions, are those which are invariant under local, i.e. gauge, transformations. We have distinguished two kinds of gauge transformations: those which act on the coordinate system of space-time, on the one hand, and those on the coordinate system of an internal space, on the other. Both are used in fundamental physics. (i) Invariance under space-time gauge transformations gives us general relativity, the classical theory of gravitational interactions. (ii) Invariance under gauge transformations of internal spaces will give us the theoretical framework to describe the other three interactions. In this

book we shall only mention gravitational interactions very briefly in the last chapter; we shall concentrate instead on the other three, the weak, the electromagnetic, and the strong interactions.

2. Spontaneous symmetry breaking for some of the internal symmetries. The necessity for this step is obvious for weak interactions whose intermediaries, the $W^{\pm}$ and the Z^0 gauge bosons, are massive. In contradistinction, the photon, which is the gauge boson of the electromagnetic interactions, is massless and the corresponding gauge symmetry should remain unbroken. We shall discuss the situation concerning strong interactions later.

6.2 The electromagnetic and weak interactions

A priori it is not at all obvious why these two interactions should be studied together. At first sight, they have very little in common, except for the fact that they both have spin one intermediaries. But this is true for every gauge theory of an internal symmetry. On the contrary, almost everything seems to separate them:

1. Weak interactions violate parity; the electromagnetic interactions conserve parity.
2. Weak interactions seem to involve only quarks and leptons of left chirality; electromagnetic interactions involve both chiralities and do not distinguish right from left. Consequently, as we explained in Chapter 4, without spontaneous symmetry breaking, weak interactions cause quarks and leptons to be massless.
3. The weak interaction intermediate gauge bosons are massive; the photon is massless.
4. Electromagnetic interactions are mediated by a single gauge boson, the photon. Weak interactions involve three: $W^{\pm}$ and Z^0. In Chapter 4, we saw that we have one gauge boson for

each independent gauge transformation. It follows that the gauge symmetry of weak interactions is larger than that of electromagnetic interactions. For the first we should use the full Yang–Mills theory.

Because of all these differences, early attempts to apply the Yang–Mills theory only addressed weak interactions. It was the American physicist Sheldon Lee Glashow who first understood, in 1960, that we obtain a richer theory by considering simultaneously weak and electromagnetic interactions in a unified framework. His work was done before that of Brout–Englert–Higgs (1964) and consequently, he did not deal with a mechanism to give masses to $W^{\pm}$ and Z^{0}. The synthesis of all these ideas was due to Steven Weinberg in 1967 and Abdus Salam in 1968 (see Figure 6.1). Their model only applied to leptons and the extension to quarks came later. We are not going to follow the historical order, naturally longer to explain; we present directly the final theory. We are going to construct it deductively by taking into account all experimental results known up to today. The construction goes through three steps:

1. The choice of gauge symmetry. We must choose the transformations which are supposed to leave invariant the dynamical equations of the theory. In Chapter 4 we saw

Figure 6.1 The Standard Model: Sheldon Lee Glashow (1932–, Nobel Prize 1979); Steven Weinberg (1933–, Nobel Prize 1979); Abdus Salam (1926–1996, Nobel Prize 1979); Glashow : S.L. Glashow archives; Weinberg : S. Weinberg archives; Salam : ICTP archives.

that there exists an exact correspondence between the number of independent gauge symmetries and that of the gauge bosons which transmit the interactions. In Table A1.3 of the elementary particles we see that, experimentally, we have four intermediate gauge bosons for the weak and electromagnetic interactions: the photon (γ) and the three bosons of weak interactions (W^+, W^-, and Z^0), where +, –, and 0 indicate their electric charges. It follows that we must postulate the invariance of the theory under four independent gauge transformations. We recall that these transformations act on the coordinate system of an internal space and do not affect at all that of our space-time.

Only knowing the number of independent transformations which leave the theory invariant does not sound like a serious constraint, but in fact, the opposite is true. We can show that this number largely determines the geometrical properties of the theory and, after the work of Yang and Mills, an important part of the dynamical equations. The underlying mathematical theory had been established during the late part of the nineteenth century by the Norwegian mathematician Sophus Lie, followed by the work of the Frenchman, Élie Cartan. It is the *theory of Lie groups and Lie algebras.* It is a very beautiful mathematical theory, too technical to present here. We give a brief summary of the results which will be of interest to us in Appendix 2. The conclusion is that, with four independent symmetry transformations, there exists only one non-trivial solution which can be visualised as follows: three rotations in a three-dimensional internal space and an additional rotation around a fourth fixed axis. Thus, the geometry of the internal space is determined. In Appendix 2 we show a more abstract and more convenient way to describe these transformations.

2. In the symmetric phase all four gauge bosons are massless. Therefore, the second step consists in creating the necessary conditions for spontaneous symmetry breaking. In practice

this implies the introduction of particles, and thus of fields, of type ρ and θ, which we presented in Chapter 5. The goal is to give masses to three gauge bosons and leave the fourth massless. The latter will be identified as being the photon. We shall have a partial symmetry breaking, like the one described in the ferromagnetic example. One out of the four symmetry transformations will remain exact and the other three will be spontaneously broken. The counting of degrees of freedom we presented in Section 5.5 shows that we need three particles of θ type, one for every gauge boson which becomes massive. If we add at least one particle of ρ type, we obtain a minimum of four zero spin degrees of freedom. It is important to realise that this reasoning gives the *minimum* number of such particles. If we add more, let us say $N > 4$, we shall end up having, in the phase with spontaneously broken symmetry, $N - 3$ physical particles. One of the items in the LHC agenda is, precisely, the search for possible additional BEH particles.

Let us try to establish a correspondence between the gauge bosons in the two phases. In the symmetric phase, in which they are all massless, we have three which correspond to the rotations in the three-dimensional internal space and one which corresponds to the rotations around the fourth axis. The naive solution would be to say that this last one is the photon and remains massless and the other three become massive and transmit the weak interactions. Glashow remarked[1] that there exists a more general solution, in which the transformation that remains exact is a rotation around an axis making an angle with respect to the initial ones. In this picture, the distinction between the electromagnetic and weak interactions is the result of spontaneous symmetry breaking. Physicists have introduced a new

[1] For the understanding of the structure of the electroweak theory he was awarded, together with Weinberg and Salam, the 1979 Physics Nobel Prize.

term to describe this phenomenon and talk about *electroweak interactions*, in order to emphasise the common origin of both. The phenomenon has precise physical consequences, which have been verified experimentally. We shall present them at the end of this chapter.

3. The third step involves the introduction of the particles which are the matter constituents. There is a lot of arbitrariness in this step. Firstly, we must decide which are the elementary constituents of matter. Even if today there is a consensus in favour of quarks and leptons, we should recall that this is a phenomenological result which could change if tomorrow we discover that quarks and/or leptons are themselves bound states of other, more 'elementary' constituents. Secondly, we do not have the slightest theoretical idea concerning the total number of these quarks and leptons. As we have pointed out already, this is one of the profound questions for which the Standard Theory offers no answer. In addition, we observe in nature that all these particles are grouped together to form three 'families', but we do not know the precise reasons that dictate such an organisation. One point seems to be important (see Appendix 1): the algebraic sum of the electric charges of all the particles in a family must vanish. This is a necessary condition for the mathematical consistency of the theory, but we shall not give the proof here. In the symmetric phase all these particles are massless and they acquire their masses in the phase with spontaneous symmetry breaking through their interactions with the BEH particles. Although we are able to adjust the strength of these interactions in order to reproduce the observed mass spectrum, we have no explanation of their precise values.

Through these three steps the theory of electroweak interactions, often called *the Standard Model,* is complete. Its agreement with experiment is spectacular. We shall present the principal results shortly.

6.3 The strong interactions

Although the study of the strong interactions is outside the scope of this book, we would like to present their main properties in this section. As we shall see, they involve new and unexpected concepts whose deep understanding remains a challenge for physicists.

Strong interactions entered the scene along with the discovery of nuclear structure. We know that nuclei are composed of protons and neutrons, which we call, collectively, *nucleons*. The observed stability of nuclear matter shows the existence of attractive forces among the nucleons which, at least at distances as large as a typical nuclear size ($\sim 10^{-15}$ m), must be stronger than the electrostatic repulsion among protons. It is this observation, combined with the estimate on the range of nuclear forces, which led Yukawa in 1935 to predict the existence of π mesons, as mediators of these new forces. The discovery of these particles, *the pions*, in cosmic rays in 1947 confirmed the prediction and opened a new discipline of fundamental physics, the study of strong interactions.

Construction of the first powerful accelerators in the 1950s brought into evidence a large number of new particles subject to the strong interactions.[2] Today we know around a hundred of these, so the question of which particles are 'elementary' became almost meaningless. *The quark model*, considered initially as a mathematical model of classification, was proposed precisely in order to bring some order to this chaos. All these hadrons were assumed to be bound states of a small number of elementary constituents which Murray Gell-Mann called *quarks*.[3]

[2] We called them *hadrons* (see Appendix 1).

[3] This word comes from a rather obscure verse in James Joyce's *Finnegan's Wake*.

Table 6.1 Examples of the quark composition of various hadrons. Baryons are bound states of three quarks and mesons of a quark–antiquark pair. We also include the example of a baryon, called Λ^0, in whose composition enters a quark of the second family, the *strange* quark *s*.

From quarks to hadrons						
Hadrons	Proton	Neutron	Meson π^+	Meson π^-	Meson π^0	Λ^0
Quarks	uud	udd	$u\bar{d}$	$\bar{u}d$	$\bar{u}u$ and $\bar{d}d$	uds

We are not going to present here the historical evolution of this concept but today the physical 'existence' of quarks as basic constituents of all hadrons is no longer a matter of controversy. It is now firmly established through a large number of experiments which have observed the presence of *hard grains* in the interior of hadrons, the high energy analogue of Rutherford's experiment which in 1911 discovered the nuclei inside atoms. Table 6.1 presents some examples of the quark composition of various hadrons.

In the table of elementary particles (Table A1.3 in Appendix 1) we see that we know today six species of quarks: *u*, *d*, *c*, *s*, *t*, and *b*. Table 5.1 shows that their masses span a wide spectrum of values, from 2.3 MeV for the lightest quark *u*, to 173,000 MeV for the heaviest, *t*. The quark *t* is the heaviest elementary particle known today.

With the arrival of quarks the very nature of strong interactions changed radically. They were no longer interactions among nucleons, but at a deeper level, those of quarks. Nuclear forces would be derivable from these fundamental interactions in the same way that forces among atoms and molecules (called *van der Waals forces*[4]) are derivable from the fundamental electromagnetic

[4] Johannes Diderik van der Waals from the Netherlands (1837–1923, Nobel Prize 1910).

interactions among charged particles. It was soon realised that the forces among quarks presented several very peculiar features:

1. They are weak at very short distances but get stronger with increasing distance. Today we have measurements which cover almost four orders of magnitude, from 10^{-15} m, the typical size of a nucleon, to 10^{-19} m, the LHC resolution. In Section 6.4 we present the experimentally observed variation and compare it with the theoretical prediction.
2. The quarks do not appear as free particles. In spite of all experimental efforts, we have not succeeded in 'breaking' a hadron and liberating its constituent quarks. We call this property *confinement*. Quarks seem to be permanently confined inside the hadrons.
3. If we compare the proton mass (~938.3 MeV) with the sum of the masses of its constituents (~9.6 MeV), we find that they differ by a factor of order 100. This means that the main part of a proton's mass is not due to quark masses. This raises two questions: the first concerns the origin of the additional mass, the second concerns the proton's stability. The proton, as a bound state of three quarks, is *much heavier* than the quarks. Usually it is the other way around. For example, the mass of the deuteron, a proton–neutron bound state, is equal to ~1875.6 MeV, therefore it is *smaller* than the sum of the mass of a proton (~938.3 MeV) and that of a neutron (~939.6 MeV). The difference is called the *binding energy* and explains the deuteron's stability. But then, what prevents the disintegration of a proton into three quarks?

At first sight it looks as though a theory fitting all these experimental constraints must have almost miraculous properties. And yet, the miracle does occur. There exists one theory, and only one, which seems to satisfy all three properties. Without entering into the technical details, we shall try to obtain it by following these properties.

1. Concerning the first, let us point out that the variation of the effective strength of an interaction with the distance is a general feature of all quantum field theories, so, by itself, this is not a problem. What is peculiar with the interaction among quarks is that this variation appears to be counterintuitive. We expect the intensity to *decrease* with the distance and go to zero at infinity, while experiments involving quarks show the opposite. Using standard techniques of quantum field theory we can study this dependence and we find that, indeed, practically all theories exhibit the behaviour expected intuitively: negligible at very large distances, their strength increases with decreasing distance. All, with one exception, Yang–Mills theories in the symmetric phase, have counterintuitive behaviour. When we compute the effective force $F_{\text{eff}}(R)$ produced by such an interaction as a function of the distance R, we find an *increasing* function which goes to zero when $R \to 0$. We call this property *asymptotic freedom* and we can formulate it as a theorem: *the only asymptotically free theories are Yang–Mills theories in the symmetric phase.*
2. The theorem of asymptotic freedom tells us that, in order to satisfy the first property of strong interactions, we must assume that they are described by a gauge invariant interaction with several independent transformations, like those proposed by Yang and Mills.[5] Furthermore, we must assume that the theory is in the symmetric phase, because the property of asymptotic freedom is lost under spontaneous symmetry breaking. But this creates a new problem: in the symmetric phase the gauge bosons are massless and the interactions are long ranged. Such a behaviour seems to be in violent contradiction with the observed short range of nuclear forces, which are supposed to be derivable from the fundamental interaction between quarks.

[5] The condition of having a Yang–Mills theory is essential. We can prove that gauge invariant theories with only one mediator, like quantum electrodynamics, are not asymptotically free.

3. The answer to this question brings us to the second property of quark interactions, that of confinement. We can visualise this as follows: consider the example of the hydrogen atom. It is a bound state of a proton and an electron. The binding force is electrostatic attraction. If we introduce an external electrostatic field, the electron will be pulled to one side and the proton to the other (see Figure 6.2(a)). If the external field is sufficiently strong, we can beat the attractive force and separate the two constituents. We say that the atom is *ionised*. Consider now the case of a π^0 meson which, as we see in Table 6.1, is a quark–antiquark pair. These particles have opposite charges so, as happened with the electron–proton system, they will be pulled in opposite directions by the external field. However, experiments show that we cannot separate them, no matter how strong the external field. It is the property of confinement (see Figure 6.2(b)). Phrased more generally, confinement postulates that all the 'elementary particles' in a Yang–Mills theory in the symmetric phase, to wit quarks, antiquarks, as well as the gauge bosons which mediate the interaction, are permanently confined inside the hadrons. We cannot observe them as free particles.

Figure 6.2 The binding forces. In (a) a proton and an electron are bound by the Coulomb force which behaves like $1/R^2$. The binding can be broken and the atom ionised. In (b) a quark–antiquark pair is bound by the strong Yang–Mills force. We do not know its precise behaviour at large distances but, if it does not decrease fast enough, the quark and the antiquark will remain confined. In (b) we represent this binding by a spring. At short distances the spring is loose and the resulting force is weak. At large distances the spring is tight and the pull increases.

We have good reasons to believe that all Yang–Mills theories exhibit this property of confinement but so far we have no rigorous proof.[6] These reasons are of two types. For the first, let us come back to Figure 6.2. The possibility of atom ionisation is a consequence of the large distance behaviour of the force $F_{ef}(R)$. The energy needed to break the atom is given by the work provided by the force in order to separate the electron from the proton and bring them far apart from each other. In mathematical terms this work is expressed by the integral $\int F_{\text{eff}}(R)dR$ from the size of the atom to infinity. The Coulomb force decreases as $1/R^2$ and the integral converges. If the force $F_{\text{eff}}(R)$ which binds the quark–antiquark pair had a much slower decrease,[7] the integral would diverge and we would need an infinite amount of energy to separate the pair. Figure 6.2(b) shows an artist's view of this phenomenon. Unfortunately, we are only able to compute the function $F_{\text{eff}}(R)$ when R is very small. We do find that it is an increasing function of R, but we do not know whether this behaviour persists at large distances. The second indication comes from numerical simulations which we perform by approximating the continuum of space-time by the points of a discrete lattice. We can prove the property of confinement on the lattice, but we cannot control the limit in which the distance between two lattice points goes to zero in order to recover the continuous space.

We shall follow the same method we used for the theory of electroweak interactions and determine the number and the properties of the transformations which will be the symmetries of the strong interactions. We shall try the most 'natural' choice

[6] The rigorous proof of this property constitutes one of *the seven problems in mathematics of the 21st century*, proclaimed in the year 2000 by the *Clay Mathematical Institute*. Each is endowed with a one million dollar prize.

[7] Any behaviour, such as a slow decrease (e.g. $1/R$), a constant, or an increase in R, would make the integral divergent.

and we shall verify it *a posteriori*, by comparing its predictions with the experimental results.

Let us consider the table of masses of quarks and leptons (Table 5.1). We know of six species u, d, c, s, t, and b with their masses, as shown in the table. In Appendix 1 we remark that each quark comes in three different types, which we call 'colours'. Therefore, it is more precise to write the quarks as u_i, d_i, etc, with the index i taking three values, 1, 2, or 3. It is convenient to group them together in a lattice form with three lines and six columns:

$$\begin{array}{cccccc} u_1 & d_1 & c_1 & s_1 & t_1 & b_1 \\ u_2 & d_2 & c_2 & s_2 & t_2 & b_2 \\ u_3 & d_3 & c_3 & s_3 & t_3 & b_3. \end{array} \tag{6.1}$$

Looking at this we observe that it admits two sets of 'natural' transformations: (i) those which mix the columns and leave the lines untouched and (ii) those which do the opposite, mix the lines and leave the columns. The first cannot form an exact symmetry because the quark masses change considerably from one column to the other. So, let us try the second. These are transformations which mix the three colours in the lattice 6.1. We may be tempted to view them as acting on the coordinate system in a three-dimensional space. However, there is a complication: in this space the coordinates are the wave functions of the three quarks. We have noticed previously that wave functions in quantum mechanics take complex values, so we must consider transformations in a complex three-dimensional space. The counting of these transformations is not difficult, but we can also refer to the results presented in Appendix 2 on the number of transformations in the Cartan classification. We find that we obtain nine independent transformations which we may split into 8 + 1. We will assume that they all correspond to exact symmetries of the theory. Let us look separately at the eight and the ninth.

The latter acts on every quark. Following Noether's theorem, this implies the existence of a conserved quantity which we can identify with the number of quarks (minus that of antiquarks).

This number must remain conserved during a reaction. Taking into account the fact that quarks are confined inside hadrons, this conserved quantum number must correspond to the baryon number we introduce in Appendix 1. With the reservations we have already expressed, all present experiments show that it is indeed a conserved quantum number. Can we conclude that this gives a gauge symmetry as that associated with electric charge conservation? The answer is no, because in the table of elementary particles there is no massless particle, the analogue of the photon, for this quantum number. It follows that this is a global symmetry, which does not generate an interaction.

The remaining eight transformations form the gauge symmetry of strong interactions. We have eight massless gauge bosons, the eight 'gluons' of Table A1.3. Like the quarks, they are also confined inside the hadrons. This property of confinement explains why nuclear forces are not of long range, in spite of gluons' zero mass. The range of nuclear forces is limited by the characteristic size of confinement, which, experimentally, is of order 1 fermi, or 10^{-15} m.

The symmetry laws determine the dynamics of interactions among quarks. By analogy with electromagnetism, we called

Figure 6.3 Quantum chromodynamics: David Gross (1941–, Nobel Prize 2004); David Politzer (1949–, Nobel Prize 2004); Frank Wilczek (1951–, Nobel Prize 2004). Gross : D. Gross archives; Politzer : D. Politzer archives; Wilczek : Photo 2007 by Kenneth C. Zirkel, Wikimedia Commons.

this theory *quantum chromodynamics.*[8] It is an asymptotically free Yang–Mills theory. In Section 6.4 we present the evolution of the effective interaction strength with distance and we compare this with the experimental results (see Figure 6.5). We see that the interaction becomes very strong at distances of order 1 fermi, in agreement with the range of nuclear forces.

A last remark before closing this section. We have just seen that quantum chromodynamics contains a characteristic distance of the order of 1 fermi. In our system of units in which $\hbar = c = 1$, this distance corresponds to an energy of the order of 200 MeV, much larger than the masses of the quarks u and d which form the nucleons. We believe that this energy scale determines a new phase transition which generates the nucleon masses. This explains why, as pointed out previously, protons and neutrons are much heavier than their constituents, the u and d quarks. Again, we have no rigorous proof of this result, only indirect evidence, both phenomenological and numerical.

6.4 The Standard Theory and experiment

In earlier sections we have developed a theoretical framework in order to describe weak, electromagnetic, and strong interactions. The essential ingredient is the invariance under local transformations, which we called *gauge transformations.* Altogether we found 12 independent transformations, four for the electroweak theory and eight for quantum chromodynamics. As a result, we have 12 gauge bosons which transmit these interactions. The symmetry is partly broken through the Brout–Englert–Higgs mechanism and three among these bosons, $W^{\pm}$ and Z^0, become massive. The other nine, the photon and the eight gluons, remain massless.

This theory has revolutionised our ideas concerning the fundamental interactions. We often say, and it is usually true, that progress in physics occurs when an unexpected experimental result

[8] From the greek word χρώμα which means 'colour'. This theory was formulated by David Gross, David Politzer and Franck Wilczek (see Figure 6.3).

contradicts the current theoretical beliefs. This forces physicists to abandon old concepts and to imagine new ones. However, this time that is not the way it happened. There was no disagreement with any experiment. Gauge theories, which introduced geometry into physics, were invented and studied for their intrinsic beauty and mathematical coherence, not for experimental reasons. The proof of this coherence was first obtained by two Dutch physicists, Gerardus 't Hooft and Martinus J.G. Veltman, who shared the 1999 Nobel Prize. A remarkable feature is that the theory has often been ahead of the experiments and confirmation of the theoretical predictions was not always immediate. We saw that in the case of the Brout–Englert–Higgs particle this confirmation took almost 50 years to come. In this section we shall give a brief account of the great experimental discoveries which consolidated the theoretical edifice known for a long time as *the Standard Model* and which promoted it to *the Standard Theory of Fundamental Interactions.*

1. In 1967, when the formulation of the electroweak model was presented, we only knew the weak interaction processes mediated by the exchange of charged gauge bosons. A typical example is neutron decay, or the decay of the μ lepton:

$$n \rightarrow p + e^- + \bar{\nu}_{(e)} \quad ; \quad \mu^- \rightarrow \nu_{(\mu)} + e^- + \bar{\nu}_{(e)}. \qquad (6.2)$$

   In a gauge theory these two reactions are produced by the exchange of a charged boson W^-; see Figure 6.4(a) and (b). The reactions mediated by the neutral gauge boson Z^0 had not yet been observed. An example is given by neutrino–proton elastic scattering,

$$\nu_{(\mu)} + p \rightarrow \nu_{(\mu)} + p, \qquad (6.3)$$

   represented in Figure 6.4(c). This reaction is difficult to observe experimentally because the neutrinos are neutral; they interact very weakly and this makes their detection

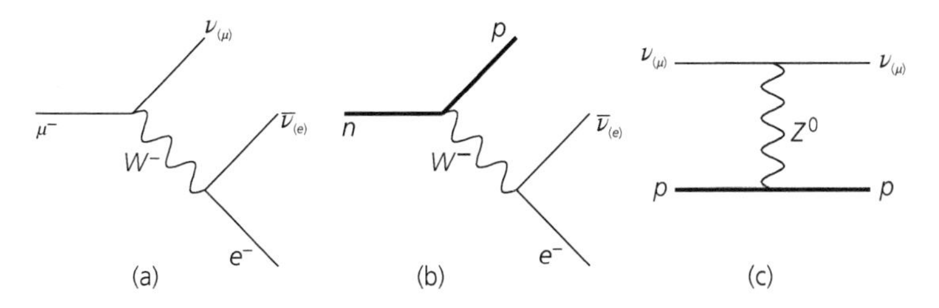

Figure 6.4 Three weak interaction processes: μ lepton decay (a) ; neutron decay (b); and elastic neutrino–proton scattering (c). The first two are produced by the exchange of a charged W, the third by that of the neutral boson Z.

very difficult. The existence of these reactions was first established at CERN in 1972, giving concrete evidence that with the gauge theories we were on the right track. The theory gives very precise predictions concerning their properties. In particular, these reactions often involve quarks and leptons with both right and left chiralities in well-defined proportions. These properties have been verified experimentally.

2. In 1967 only the first three quarks were known, u, d, and s. On the other hand we knew the four leptons of the first two families, namely the electron (e^-), its associated neutrino ($\nu_{(e)}$), the muon (μ^-), and its neutrino ($\nu_{(\mu)}$). Looking at Table A1.3, we see that the second family was not complete. The quark c was missing. This is the reason why the 1967 model proposed by Weinberg applied only to leptons. Its extension to hadrons led to the prediction of the fourth quark c (c standing for 'charm') and the new hadrons in the composition of which this quark participates. The discovery of these 'charmed' particles between 1974 and 1976 was the second great success of this theory.
3. The most characteristic prediction of gauge invariant theories is the existence of gauge bosons which transmit these interactions. As we have already pointed out, their number and their properties are fixed by the theory. For electroweak

interactions, spontaneous symmetry breaking also determines their masses. Their actual discovery was a great experimental challenge, because the predicted masses were putting them out of reach of the 1970 accelerators. A new form of proton–antiproton collider was invented in order to produce and identify them. This discovery, made at CERN in 1983, gave the 1984 Nobel Prize to Carlo Rubbia, the experimentalist who proposed and supported the project, and to Simon van der Meer, the engineer who designed and built the essential element of the accelerator. Later, in the 1990s, the properties of the W and Z bosons were studied in detail and all theoretical predictions precisely confirmed.

4. We point out in Appendix 1 that the mathematical consistency of the electroweak theory requires the families to be complete. We cannot have a doublet of leptons without the corresponding quarks. More precisely the theorem states that the algebraic sum of the electric charges of all particles in a family (quarks + leptons) must vanish. The proof is quite technical and we shall not present it here. But this theorem yields precise predictions. The discovery of a new lepton, the τ, together with its associated neutrino, at the Stanford Linear Accelerator Centre (SLAC) by Martin Lewis Perl[9] was interpreted as the opening of a new family and the prediction of two new quarks. Indeed, the quark b was discovered by Leon Lederman at FermiLab, near Chicago, in 1977, and the quark t a little later, first indirectly at CERN and then directly at FermiLab.
5. The success of quantum chromodynamics is based on the property of asymptotic freedom which predicts a weak coupling at short distances. Figure 6.5 shows a theoretical prediction of the variation of the effective coupling with the distance, together with the experimental points. The agreement is impressive.

[9] 1995 Nobel Prize.

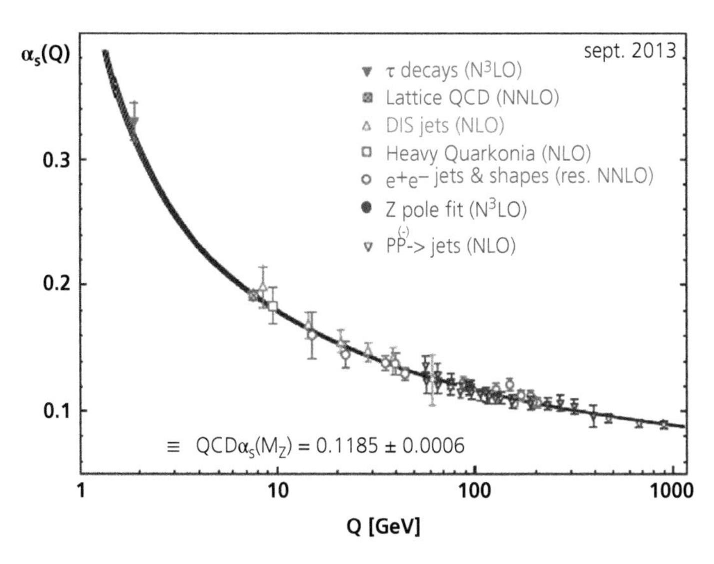

Figure 6.5 The variation of strong interaction effective strength with the energy scale. The points with error bars are the experimental, or numerical, measurements. The width of the curve represents the theoretical uncertainties. Notice that, in our system of units, distance is the inverse of energy. CERN, CMS collaboration.

6. According to quantum chromodynamics, the strongly interacting 'elementary particles' are the quarks and the gluons. The latter are those which transmit the interaction. Therefore, we must find these inside the hadrons, together with the quarks. It is more difficult to detect them because we use mostly electromagnetic interactions to probe the interior of hadrons, and the gluons are electrically neutral. Nevertheless, they have been detected and identified at the DESY research centre in Hamburg.
7. The curve of Figure 6.5 shows that at distances of the order of a few fermi (or, approximatively, 1 GeV) the interaction becomes very strong. It is the region in which the hadronic bound states, such as the proton, the neutron, the mesons etc., are formed. In Section 6.3 we explained that a phase transition occurs which generates the largest part of the

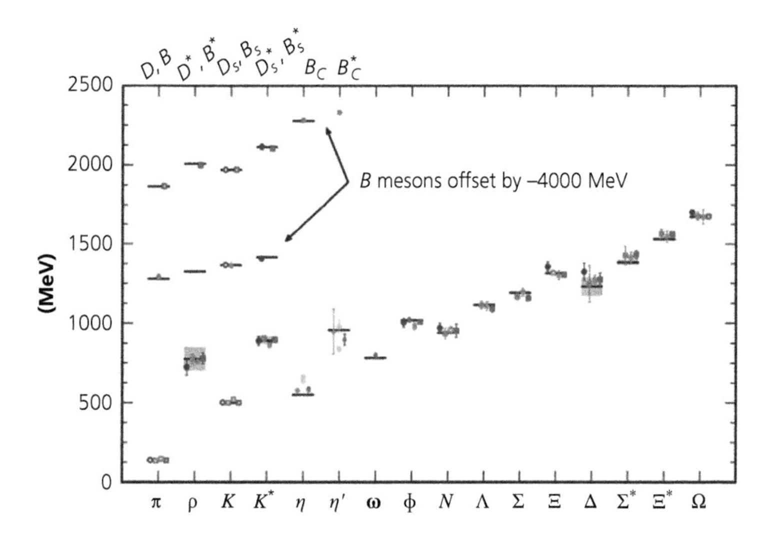

Figure 6.6 The mass spectrum of the light hadrons computed using numerical simulations of quantum chromodynamics on a space-time lattice. The masses of hadrons with a *b* quark are displaced by 4 GeV. The figure also shows the uncertainties, both theoretical and experimental. The agreement is impressive. A. Kornfeld, ArXiV 1203.1204.

mass of these hadrons. Our ability to perform analytical calculations is limited to the weak coupling regime, so, in order to study these strong interaction phenomena, we appeal to numerical simulations. Figure 6.6 shows the results of such simulations for the lightest hadronic states, together with the experimental points. Another confirmation of the Standard Theory.

8. *Last but not least,* a brilliant confirmation of the theory is the recent discovery of the Brout–Englert–Higgs particle, predicted half a century ago.

7

Epilogue

With the discovery of the Brout–Englert–Higgs particle, the trace of mass generation, the Standard Theory is complete. All its predictions have been brilliantly confirmed by experiment. Is this the end of the story? In this chapter we shall argue that the answer is no. We shall indicate possible avenues to go beyond the Standard Theory towards a richer and more coherent model.

The guiding principle is that the precise knowledge of physics at a given energy scale allows us to guess the possible presence of new physics at a higher scale. An analogy: imagine that extraterrestrials are observing us from outer space with telescopes whose resolution power is limited to 10 m. They can distinguish our constructions and large pieces of equipment of all kinds, buildings, bridges, trains, ships, planes, etc., but they cannot see and study us. However, based on their observations, they can easily deduce the presence of living creatures on this planet and even guess some of our properties: our approximate size must be of the order of 1 m, (all these constructions are not the work of ants), that we live in open air, etc. They can therefore conclude that, by increasing the resolution power of their telescopes, they may make a discovery, that of new inhabitants of the Universe.

In physics we often find ourselves in a similar situation. The knowledge of the range of nuclear forces led Yukawa to predict the existence of the π meson. The properties of weak interactions suggested the existence of a new class of hadrons, the charmed particles. We have good reasons to expect that the story will repeat itself and the detailed study of the Standard Theory will reveal to us the existence of new physics, hopefully accessible to

LHC. We want to present here very briefly some of these reasons. They are of two sorts:

1. We have already mentioned several questions to which the Standard Theory offers no answer. It does not explain why quarks and leptons appear in three families, neither why their masses are spread over at least 11 orders of magnitude.[1] The very origin of the neutrino masses remains a mystery. In the same spirit, in the Standard Theory the observed electric charge quantisation, i.e. the fact that electric charges of all particles appear to be integer multiplets of an elementary charge, is a mere coincidence.[2] A truly fundamental theory should be able to answer these types of questions.
   The most important limitation of the Standard Theory is the fact that it ignores gravitational interactions. This does not affect its agreement with experiment because gravitational effects are completely negligible in particle physics experiments. However, this omission shows that the theoretical model is incomplete. This is even more so, since, as we remarked in Chapter 3, the gravitational interactions are described classically by general relativity, which is the gauge theory of space-time symmetries. In spite of all the theorists' efforts and ingenuity, we have not been able to incorporate gravitation into the Standard Theory. In fact we are essentially convinced that quantum field theory may not be the right framework to describe quantum gravity. During the past 30 years efforts have moved away from the concepts of point particles and local fields towards the more general ones of extended objects, strings, or membranes. From the theoretical point of view, this approach has produced

[1] We recall that the top quark mass equals 173 GeV ($1\,\text{GeV} = 10^9\,\text{eV}$) and the neutrino masses are smaller than 1 eV.

[2] A large piece of matter, containing a huge number of protons and electrons, always appears electrically neutral. This puts a very stringent limit on a possible difference between the absolute values of the electric charges on these two particles. The limit can be written as: $|Q_P/Q_e| = 1 \pm \epsilon$ with $\epsilon < 10^{-20}$.

models of great mathematical depth. It offers the only consistent framework to bring together the two major discoveries of the early twentieth century, general relativity and quantum mechanics. On the other hand, from the physical point of view, we are still far from obtaining models with sufficient predictive power in order to compare them with experiment.

2. One could object that all the limitations mentioned earlier only show that the Standard Theory is not *The final theory* of all interactions, hardly a surprising conclusion from the epistemological point of view. They predict the existence of new physics, but don't allow us to estimate the energy scale of its appearance. In the absence of such an estimation, even as an order of magnitude, the prediction is not very interesting. In other words, a prediction must be quantitative and not merely qualitative. We would like to argue here that the Standard Theory is sufficiently precise to allow for quantitative predictions.

   Let us come back to the Standard Theory as exposed in Chapter 6. It is a gauge invariant theory. In the low temperature phase, the phase of our present Universe, the symmetry is partly broken and the gauge bosons $W^{\pm}$ and Z^0 which mediate the weak interactions, as well as the fermions which are the building blocks of matter, are massive. They acquired their masses through the Brout–Englert–Higgs mechanism. In the high temperature phase, the symmetric phase, all these particles are massless; all but the four BEH spin zero bosons. They are the only particles which do not get their masses through the BEH mechanism. The mass of these bosons in the symmetric phase is the only dimensionful parameter of the theory, the one which determines the mass scale. We have seen that, experimentally, this scale is given by the BEH boson mass, which equals 126 GeV.

   The theory contains in addition other dimensionful parameters at higher scales. For example, the gravitational interactions introduce Newton's constant, which, expressed in GeV, corresponds to a huge scale of order 10^{19} GeV, often called *Planck's mass*, M_{Pl}.

The simultaneous presence of two so widely separated mass scales is not natural in a fundamental theory. It implies the introduction of a dimensionless constant, their ratio, which is of order 10^{-17}.[3] We cannot believe that a fundamental theory contains parameters of that order. In fact, the situation is even more serious: quantum field theory allows us to estimate the corrections to the BEH mass induced by the presence of the Planck mass. Not surprisingly, these corrections are proportional to M_{Pl}^2. In other words, the theory is not able to naturally sustain two mass scales, so far away from each other. Strictly speaking the problem is aesthetic: we do not like the presence of parameters equal to 10^{-34} in equations of a fundamental theory. Nevertheless, experience has taught us that aesthetic criteria are often good guides for deciphering nature's secrets.

Is there a solution to this problem? The answer is yes, and even more than one, but they all imply new physics, often at a scale immediately above that of the BEH particle, of the order of 1000 Gev. The details of this new physics are model dependent and cannot be stated precisely. We are in the place of the extraterrestrials who are looking at our planet. They know that it is inhabited, but they cannot guess the details of our appearance. They do not know whether we are four-legged, two-legged, or three-legged. The new physics we speculate about, often contains new 'elementary' particles with masses of order 1000 GeV, but their nature and their properties depend on the particular assumptions we make.

A model which has been analysed in detail postulates the existence of a new symmetry which relates fermions and

[3] In fact, the parameters which enter in the fundamental equations are the squares of the masses, so the dimensionless parameter is of order 10^{-34}.

bosons. We call it *super-symmetry* and it has remarkable mathematical properties. We also find it as an important ingredient of string theories which attempt to build a consistent theory of quantum gravity. Supersymmetry predicts the existence of a whole series of new particles, the supersymmetric partners of all existing particles.

There are even more exotic ideas according to which, at a scale of 1000 GeV, the number of space dimensions changes and we may discover hidden ones. An example: imagine a cylinder of radius r and length l, much larger than r. If we look at it with a spatial resolution worse than r we will think we are looking at an one-dimensional line. With better resolution we will find out it is in fact a two-dimensional cylindrical surface.

LHC has already given us the BEH particle. In 2015 it started operating again at higher energies. Naturally, the first item on its agenda is the detailed study of the new boson properties. We have very precise theoretical predictions about these. For example, its interactions with quarks and leptons must depend crucially on their masses, since it is through these interactions that these particles become massive. All this must be measured very accurately if we want to confirm all aspects of the BEH mechanism. But, in addition to this, already very rich programme, it is the search for new physics which fascinates physicists. The two ideas we have mentioned, supersymmetry and hidden dimensions, are two examples among many others. Nature may be hiding further surprises. We are reasonably confident that new physics lies at higher energies and that LHC will be able to find it. Great discoveries, which seem to mark the end of a story, mark in fact the beginning of a new, more extraordinary one.

Appendix 1

The Elementary Particles

A1.1 Introduction

If we divide a drop of water into two parts we obtain two drops. They are smaller but the substance is the same. If we repeat the exercise and divide each one into two, we obtain four. Carrying on we obtain successively eight, sixteen, etc. droplets. How many times can we continue? Can we envisage that this process will lead us eventually to the smallest possible drop of water, *the elementary drop*, or, contrary to this, that it is an endless process? In other words, is the structure of matter continuous, or discontinuous?

It seems that it was Democritus from Abdera, Greek philosopher from the end of the fifth century BC,[1] who first gave the right answer: *discontinuous* (see Figure A1.1). Although we do not know any details of his reasoning, we do know that his answer was correct. There does exist an 'elementary drop' of water, which we call a *water molecule*, and we know very well its properties and its chemical composition. Democritus considered these elementary constituents of matter as 'unbreakables' (he called them 'atoms'), but here he was only half right. We can break a water molecule, but the pieces are no longer water.

Today, matter composition is no longer a philosophical question, but a domain of intense experimental research. This evolution is due to the technical progress which has made possible the design, construction,

[1] An almost legendary figure, Democritus was born probably in Abdera of Thrace at around 460 BC and died in 370. Although he was essentially a contemporary of Socrates, he is considered among the pre-Socratic philosophers. According to ancient sources, his work had been immense, but only small fragments of it are known to us. On the other hand numerous apocryphal stories are attributed to him. Together with his master Leucippus, he is considered to be the father of atomic theory. According to Democritus, matter is composed of atoms, ('impossible to break', from the Greek word τέμνω, which means 'to break'), and the *vacuum* which fills the space between the atoms.

Figure A1.1 The 'fathers' of the structure of matter: Democritus (460–370 BC), the inventor of the atomic concept; Ernest Rutherford, 1st Baron of Nelson (1871–1937, Nobel Prize in Chemistry 1908) the first atomic model; Murray Gell-Mann (1929–, Nobel Prize 1969), the quarks.

Table A1.1 The resolution power of the principal microscopes.

Microscopes	
Naked eye	10^{-4}–10^{-5} m
Optical microscopes	10^{-7} m
Electronic microscopes	10^{-10} m
X rays	10^{-11} m
α rays	10^{-13} m
Accelerators $\sim$ 100 MeV	10^{-14}–10^{-15} m
Accelerators $\sim$ 10 GeV	10^{-16}–10^{-17} m
L.E.P., Tevatron	10^{-18} m
LHC	10^{-19} m

and operation of more and more powerful microscopes. Table A1.1 gives a list of the most important types.

Some words of caution concerning this table. The indicated resolution powers are only approximate because, in fact, the table contains instruments which are not directly comparable. The resolution of the human eye is limited by physiological factors, that of optical microscopes by the wavelength of visible light, of order 5×10^{-7} m. A first revolution came in 1895 with Wilhelm Röntgen, (1845–1923, first Physics Nobel Prize in 1901), who discovered X rays with wavelength 10^{-8} m and obtained his famous pictures. The motivation was not so much resolution, but rather penetration. With X rays we can 'see' bones and other internal organs. Since they are not in the domain of visible light, it is not with our eyes that we see the image, but on a photographic plate, or a computer screen. Today the X ray technique reaches wavelengths of order 10^{-11} m or shorter, for the study of structures in materials, or biological macro-molecules. Figure A1.2 shows the spectrum of electromagnetic radiation. The separations between γ rays, X rays, etc. are conventional. Visible light occupies only a small part of the spectrum, between 400 and 750 nm.

A second revolution was due to Ernest Rutherford, 1st Baron of Nelson, who, in 1909, had the idea to use α particles. Along with Hans Geiger and Ernest Marsden, he studied the scattering of α particles, in fact helium nuclei, from a thin foil of gold. The results were astonishing. Most α particles passed through unaffected, but, occasionally, some were deflected at large angles. Rutherford interpreted these results as meaning that the space occupied by the atoms was mostly empty with some hard grains inside. He proposed a classical atomic structure following the stellar model: a massive positively charged nucleus in the middle with electrons, with light masses and negative electric

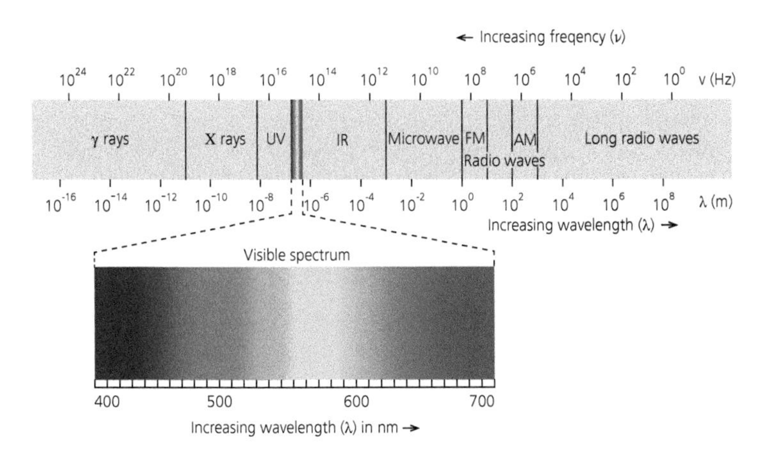

Figure A1.2 The spectrum of electromagnetic radiation. The part containing visible light is shown expanded. The figure shows also some of the principal applications of electromagnetic waves according to their wavelength.

Figure A1.3 The first fathers of quantum theory: Max Karl Ernst Ludwig Planck (1858–1947, Nobel Prize 1918) the quantum concept; Niels Henrik David Bohr (1885–1962, Nobel Prize 1922) the old quantum theory; Prince, later Duke Louis-Victor Pierre Raymond de Broglie (1892–1987, Nobel Prize 1929) the wave–particle duality. We could add Albert Einstein for the photoelectric effect; see Figure 2.1.

charge, gravitating around it. Today we know that this simple model is very naive: a classical electron orbiting around the nucleus would lose energy by radiation and would fall rapidly into the centre. It was the Dane, Niels Bohr (Figure A1.3) who completed the picture with the introduction of quantisation rules which later gave rise to quantum mechanics.

Rutherford, by introducing with his model the concept of the atomic nucleus, became the founder of nuclear physics, but in addition, with his revolutionary experimental method, the father of a new generation of 'microscopes', which use energetic particles to probe the structure of matter. Table A1.1 shows the evolution of this method. The most powerful microscopes are, in fact, the particle accelerators. The natural quantity to label their power is the final energy of the accelerated particles. We express this in electronvolts (eV), but we often use multiples of this unit: 1 MeV $= 10^6$ eV and 1 GeV $= 10^9$ eV. We recall that in our system of units in which $c = \hbar = 1$, energy has dimensions: $[\text{energy}] = [\text{distance}]^{-1}$. Substituting the numerical values of c and $\hbar$, we find formula (1.1), which gives $1\ \text{f} = 10^{-15}\ \text{m} = [200\ \text{MeV}]^{-1}$. This allows us to find an approximate relation between the energy of an accelerator and its spatial resolution. Table A1.1 shows that in the course of the twentieth century we gained a factor of 1 million in resolution. Using these instruments the following were discovered successively:

molecules $\rightarrow$ *atoms* $\rightarrow$ *nuclei* + *electrons* $\rightarrow$ *protons* + *neutrons* + *electrons* $\rightarrow$ *quarks* + *electrons* $\rightarrow$ *???*

Figure A1.4 presents 'an artist's view' of this process towards the infinitely small.

There is no reason to think that the series ends somewhere, and even less, that we have already reached this end.

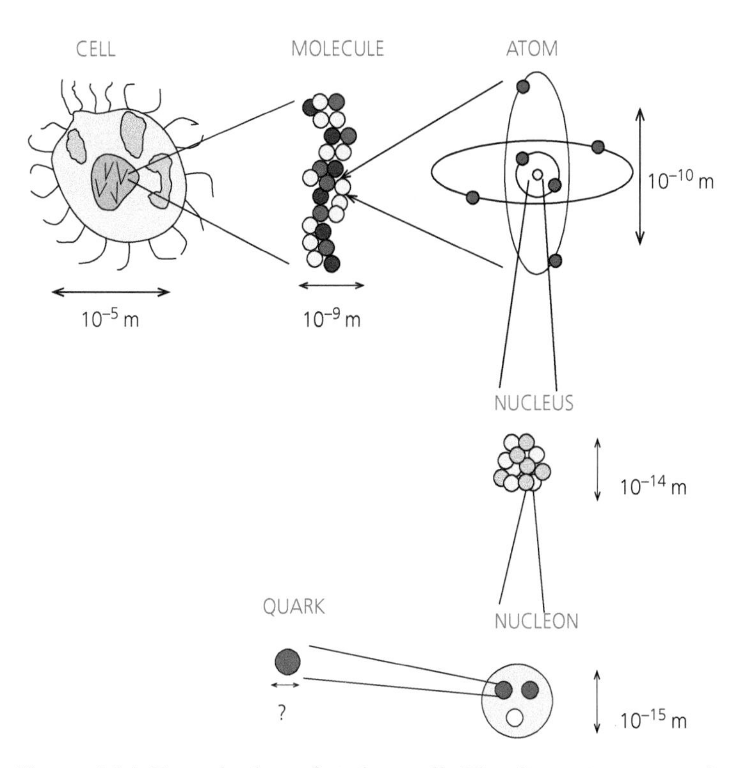

Figure A1.4 Towards the infinitely small. The dimensions are only indicative, especially for cells and biological macromolecules. In fact these dimensions vary considerably according to the cell, or the molecule. For the quarks we know only that they are smaller than $\sim 10^{-18}$m–10^{-19}m, which is the LHC resolution.

A1.2 The four interactions

Nature is characterised by an incredible diversity. The variety of forms, structures, colours, and properties which are offered to our senses seems to be infinite. Nevertheless, we know today that this is only apparent; nature is indeed plethoric in forms and properties, but it is extremely economical in building blocks and fundamental forces. If, as we have just seen, our ideas on the elementary constituents of matter have profoundly evolved during the last century, those on the fundamental forces have remained remarkably stable. No matter which scale in matter composition we are looking at, from the smallest particles we produce with our accelerators, to the largest clusters of galaxies in the Universe, as well as all degrees of complexity, from the simplest hydrogen atom to the most complex biological molecule, all owe their structure to four fundamental forces. In increasing strength, they are the following:

1. *Gravitational interactions.* These are well known from our everyday experience. They are responsible for falling apples, for the motion of the planets around the sun, and the structure of the Universe. However, at the microscopic scale, their intensity is negligible, much weaker than anything we have been able to measure up to now.
2. *Weak interactions.* These are responsible for most radioactive decays and they initiate the fusion processes at the origin of stellar energy.
3. *Electromagnetic interactions.* These are also well known in macroscopic physics. They are repulsive between same charged particles and attractive between oppositely charged ones. They have a long range and they are responsible for the structure of atoms and molecules, as well as many properties of condensed matter.
4. *Strong interactions.* These are responsible for nuclear structure. They are attractive between the constituents of nuclei (protons and neutrons) and stronger than the electrostatic repulsion between protons. They have a short range of order $1\,\mathrm{f} = 10^{-15}$ m.

A large part of this book is devoted to the theoretical description of these interactions. From a historical point of view, it is a process which started long ago, but the modern version of theoretical particle physics has a precise birth date: 2 June 1947, the date of the Shelter Island Conference. The most important contributions presented in this conference were not brilliant theoretical breakthroughs, but two experimental results. Willis Eugene Lamb (1913–2008, Nobel Prize 1955), of Columbia University, announced the measurement of an energy shift between the levels $2^2S_{1/2}$ and $2^2P_{1/2}$ of the hydrogen atom and Isidor Isaac Rabi (1898–1988, Nobel Prize 1944), that of an 'anomalous' magnetic moment of the electron. The interest in these results was that, for the first time, they were in clear contradiction with the Dirac (Figure A1.5) theory, which was 'the bible' of theoretical physics up to that time. They forced theorists to elaborate a more general and mathematically coherent theory which revolutionised our fundamental physical ideas and gave rise to the theory we describe in this book.

This theory is called *quantum field theory* and it is the quantum mechanics of a system with a very large, in fact infinite, number of degrees of freedom. Attempts to formulate such a theory go back to 1925, with the work of many physicists among whom we find Born, Heisenberg, Jordan, Dirac, Pauli, Fermi, etc., but, under its modern form, it was developed by Richard Phillips Feynman, Julian Seymour Schwinger,

Figure A1.5 Together with W. Heisenberg, see Figure 3.8, they are the fathers of quantum mechanics: Erwin Rudolf Josef Alexander Schrödinger (1887–1961, Nobel Prize 1933) the first wave equation in quantum mechanics; Paul Adrien Maurice Dirac (1902–84, Nobel Prize 1933) the relativistic wave equation and the first quantum field theory computation.

Figure A1.6 The founders of 'modern' quantum electrodynamics: Richard Philips Feynman (1918–88, Nobel Prize 1965); Julian Seymour Schwinger (1918–94, Nobel Prize 1965); Sin Itiro Tomonaga (1906–79, Nobel Prize 1965).

and Sin Itiro Tomonaga (Figure A1.6) in 1947 and completed the following year by Freeman John Dyson. The first application is named *quantum electrodynamics* and describes with high precision the interaction of electrons with an electromagnetic field. An example of the extraordinary agreement between this theory and experiment is given by the *electron anomalous magnetic moment* measured by Rabi.[2] The most recent values are:

$$\begin{aligned} a_e(exp) &= 0.00115965218076(27) \\ a_e(th) &= 0.00115965218178(77), \end{aligned} \tag{A1.1}$$

[2] The magnetic moment is the quantity which determines the interaction of a particle with a magnetic field. That of the electron, denoted as μ_e, is usually expressed in units of 'an elementary magnetic moment', *the Bohr magneton*, given by $\mu_B = e\hbar/2m_e$, where e and m_e are the electron charge and mass, respectively. The term *anomalous* has a purely historical origin. Before the advent of quantum electrodynamics, physicists were computing the electron magnetic moment using Dirac's equation, which is the relativistic generalisation of the quantum mechanical Schrödinger equation for the electron. The Dirac equation predicts a value for the ratio $g_e = \mu_e/\mu_B$, called the *gyromagnetic ratio*, equal to 2. The early experiments, which were not very precise, were compatible with this prediction and provided one of the successes of the Dirac theory. Therefore, a value different from 2, like that announced by Rabi, was at first considered to be 'anomalous'. Of course, today we know that there is nothing anomalous about such a value, but the term has survived and the results are always presented for the quantity $a_e = (g_e - 2)/2$.

where we have indicated the theoretical prediction of quantum electrodynamics. The numbers in parentheses correspond to the uncertainties, experimental or theoretical. The agreement is impressive. In fact, the precision of this measurement, combined with the theoretical calculations, yield today the best determination of the value of the electron charge.

A remarkable feature of quantum electrodynamics is that it associates a particle with the electromagnetic field, the 'quantum' of the field, which we call a *photon*.[3] The interaction between charged particles is described by the exchange of one or more photons. We say that the photon is *the messenger* of the electromagnetic interactions, a notion which, as we saw in Chapter 4, is generalised to all other interactions. By analogy, we shall call the messengers of all interactions *radiation quanta*.

A1.3 Some basic notions

Before presenting the table of elementary particles, we want to introduce a dictionary of some concepts and the associated terminology, which will be useful throughout this book. They are the various properties which we associate with elementary particles.

1. *Mass*. This is the familiar concept which we know from classical mechanics: under the influence of a force $\vec{F}$, the particle acquires an acceleration $\vec{a}$, given by Newton's formula $\vec{F} = m\vec{a}$. A first abstraction comes from special relativity, which tells us that the value of the mass that enters Newton's formula depends on the speed of the particle and is given by $m = m_0/\sqrt{1 - v^2/c^2}$, where m_0 is the mass of the particle in the coordinate system in which it is at rest and c is the speed of light in vacuum. It is the mass m and not m_0 which we use in Einstein's formula $E = mc^2$. It follows that a massive particle can never reach the speed of light, because the resulting energy would be infinite. On the other hand we can have particles whose 'rest mass' is equal to zero but which have non-zero energy because their speed equals the speed of light

[3] The concept of the photon as the mediator of an electromagnetic interaction had been introduced earlier, but it is in the framework of quantum electrodynamics that this concept acquires a precise meaning.

in all reference frames. Such particles exist in nature, the well-known example being the photon, the particle associated with an electromagnetic field.

An old theoretical idea is that particles acquire their mass, at least partly, through their interactions. Already by the end of the nineteenth century Lorentz had attempted to compute the 'self energy' of the electron, i.e. the energy coming from the electron's interaction with its own electromagnetic field. Since the electrostatic potential of a point charge at a distance r is proportional to $1/r$, this energy, computed at the point $r = 0$ of the position of the particle, is, in fact, infinite. This problem haunted theoretical physics throughout the first half of the twentieth century and it was only with the advent of quantum field theory that we understood that this particular problem is due to the fact that the question is not well defined. In the electron example, we can only measure its physical mass and not that which it would have had in the absence of its own field. However, the question, appropriately rephrased, is still relevant and people have tried, though unsuccessfully, to attribute the small proton–neutron mass difference to the proton's electric charge.

2. *Quantum numbers.* This concept generalises that of electric charge. A particle's charge can be positive, negative, or zero. We know that electric charge is a conserved quantity: in a reaction, the algebraic sums of all electric charges in the initial and the final states are equal. Charge conservation has an immediate consequence: the electron, which is the lightest charged particle, is stable. All observations reveal another remarkable property of electric charge: all charges are integer multiplets of an elementary unit of charge. We do not have a deep understanding of this property and this shows that our theoretical ideas are still incomplete.

   Experiments indicate the existence of additional quantities which appear also to be conserved. We call these 'quantum numbers' and they also take discrete values. The most common example is the 'baryon number'[4] which equals 1 for the constituents of nuclei, the proton and the neutron,[5] and zero for electrons. The study of nuclear reactions shows that, although protons and neutrons may transform into each other, their absolute number

[4] From the Greek word βαρύς, which means 'heavy'.

[5] We call them collectively *nucleons.*

remains constant. We have never observed nucleons appearing, or disappearing spontaneously. If baryon number is conserved, the proton, which is the lightest baryon, must be stable and this would explain the great stability of nuclear matter. On the other hand, if this conservation is exact, we cannot explain the creation of baryons just after the Big Bang. For this reason physicists consider that baryon number can be only conserved approximately and considerable experimental effort aims at detecting a possible proton decay.

Another remark concerning quantum numbers. Relativistic quantum mechanics predicts, and experiments confirm, that for every particle there is an associated *anti-particle*, which has exactly the same mass as the particle, but carries the opposites of all quantum numbers. To the electron there corresponds the *positron*, which has a positive electric charge,[6] to the proton, the *antiproton* with negative electric charge and baryon number, to the neutron, the *antineutron* with zero electric charge and baryon number equal to −1, etc. There exist particles, like the photon, which carry no quantum number different from zero. It follows that they are indistinguishable from their anti-particles. The 'antiphotons' are identical to the photons.

3. *Spin*. In classical mechanics this term denotes the motion of a spinning top and it is a special case of the more general concept of *angular momentum*. In quantum mechanics we define a particle's spin as its intrinsic angular momentum, but we must abandon the image of a spinning top, which is meaningless for a point particle. Like many other quantum mechanical quantities, the spin can take only discrete values, which, in the appropriate units, are positive numbers, integers, or half-integers. Particles with integer spin (0, 1, 2, ...) are called *bosons*, in honour of the Indian physicist Satyendra Nath Bose (1894–1974) and those with half-integer spin (1/2, 3/2, ...) *fermions*, in honour of the Italian physicist Enrico Fermi.[7] The photon, which has spin equal to 1, and the BEH

[6] The existence of an electron's anti-particle was predicted by Dirac as a consequence of the equation of relativistic quantum mechanics which bears his name. This prediction was brilliantly confirmed in 1932 by Carl David Anderson (1905–91, Nobel Prize 1936), who discovered the positron in cosmic rays.

[7] To be precise, this result, along with many others we present in this book, applies to particles in our three-dimensional space. This precision is necessary

particle, which has spin equal to 0, are examples of bosons,[8] while the electron, the proton, the neutron, etc., which have spin equal to 1/2, are examples of fermions.

In our three-dimensional space there is a profound difference in the behaviour of these two kinds of particles. Let us consider two identical particles, one at point $\vec{x}$ and the other at point $\vec{y}$. In Box 3.1 we explained that in quantum mechanics, the state of a physical system is described by a complex valued function which we call a 'wave function'. We also explained that the phase of this function has no physical significance. It follows that the state of these two particles will be described by a wave function $\Psi(\vec{x}, \vec{y}, t)$ defined up to a phase. Let us imagine now that we interchange the two particles and we bring the first one to point $\vec{y}$ and the second to $\vec{x}$. Since the particles are assumed to be identical, we expect the new system to be described by the same wave function, up to a phase:

$$\Psi(\vec{x}, \vec{y}, t) = C\Psi(\vec{y}, \vec{x}, t), \tag{A1.2}$$

where C is a complex number of modulus 1. If we repeat the exchange operation we bring the system back to its original state, but the wave function is multiplied by C^2,

$$\Psi(\vec{x}, \vec{y}, t) = C\Psi(\vec{y}, \vec{x}, t) = C^2\Psi(\vec{x}, \vec{y}, t). \tag{A1.3}$$

We conclude that $C^2 = 1$ and, consequently, $C = \pm 1$. The wave function of two identical particles is either *symmetric*, or *antisymmetric*, under their exchange. It turns out that a deep theorem of quantum field theory proves, and experiment confirms,

because the properties of a physical system often depend on the number of dimensions of the ambient space. In particular, the spin concept, as well as the distinction between fermions and bosons, change if we consider a system on a two-dimensional surface, or a one-dimensional line. This is easy to understand. We defined spin as an angular momentum, therefore it is related to the properties of spatial rotations. These properties depend on the number of space-dimensions. In a three-dimensional space we can perform three independent rotations, on a plane we have only one, and in a line the very notion of rotations is lost.

[8] We often call it the 'BEH boson'.

that under such an exchange, the wave function is *symmetric*, $\Psi(\vec{x}, \vec{y}, t) = \Psi(\vec{y}, \vec{x}, t)$, if the particles have integer spin (bosons), and *anti-symmetric*, $\Psi(\vec{x}, \vec{y}, t) = -\Psi(\vec{y}, \vec{x}, t)$, if they have half-integer spin (fermions). This implies in particular that for the fermions, the wave function must vanish if $\vec{x} = \vec{y}$; in other words, two fermions cannot occupy the same place.[9] This difference, which sounds purely technical, is at the origin of the partition of electrons in atomic orbits giving rise to the structure of matter. Let us take, for example, an atom with N electrons. Because of Pauli's principle, they cannot all sit at the lowest energy state and they are forced to occupy successive atomic shells, whose number increases with N. This is the explanation for Mendeleev's periodic table of elements. In contradistinction, there is no exclusion principle for bosons and nothing prevents an arbitrarily large number of them agglomerating in the same quantum state. We call this phenomenon *Bose–Einstein condensation* and we expect to find it at low temperature where thermal agitation is small. Many physical phenomena with important technological applications result from this symmetry property of bosons. To name but a few, we have the laser, and also superconductivity, superfluidity, etc. In Chapter 5 we indicated that the electroweak phase transition, the subject of this book, which is responsible for the creation of most particle masses, is precisely due to the bosonic character of the Brout–Englert–Higgs particle.

4. *Leptons, hadrons*. In Section A1.2 we introduced the four fundamental interactions between elementary particles. Experiments show that some particles are not subject to all these interactions. In particular, some fermions, like the electron, are insensitive to strong interactions and we call them *leptons*.[10] Others, such as the nucleons, protons, and neutrons, are sensitive and we call them *hadrons*.[11] In the way we introduced the baryon quantum number, we can also introduce a *leptonic quantum number*, or *lepton number*. This

[9] This property was first postulated by W. Pauli for electrons and it is known as *Pauli's exclusion principle*. To be precise, it applies under the complete exchange of the two particles, including their position in space as well as other variables which characterise their state, such as their spin orientations.

[10] From the Greek word λεπτός, which means 'fine' or 'thin'.

[11] From the Greek word αδρός, which means 'strong'.

equals 0 for the hadrons and 1 for the leptons; the electron has leptonic number equal to +1 and its anti-particle, the positron, −1.

A1.4 The neutrino saga

The neutrino is the most elusive and probably the most fascinating of all elementary particles. Its story is an extraordinary and instructive chapter in the history of modern physics. It is intimately related to all developments of the twentieth century, it involves remarkable experiments and brilliant theoretical ideas, but also it provides for an exemplary illustration of the scientific process. We shall present here a very short account of the early part of the story.

It begins with the discovery of nuclear β-decay by Henri Becquerel[12] in 1896 and the identification of the emitted particles as electrons in 1902. Today we know that the reaction is, in fact, the decay of a neutron giving a proton, an electron, and an anti-neutrino. For a neutron bound to a nucleus, this reaction appears as a nucleus giving a second nucleus, an electron, and an anti-neutrino:

$$n \rightarrow p + e^- + \bar{\nu} \Rightarrow N_1 \rightarrow N_2 + e^- + \bar{\nu}. \tag{A1.4}$$

For the physicists of the early twentieth century this reaction was a seemingly endless source of questions:

1. They knew neither the neutron nor the neutrino. In any case, with the technology available at the time, the latter was undetectable. So, they were seeing only

$$N_1 \rightarrow N_2 + e^-. \tag{A1.5}$$

2. Since they could not imagine that an electron could come out of a nucleus without being there in the first place (the quantum field theory for electrons which describes how particles can be created by the interaction was first formulated by E. Fermi in 1933), they had a nuclear model according to which nuclei were bound states of protons and electrons.

[12] Antoine Henri Becquerel, 1852–1908, Nobel Prize 1903.

But the serious problems started with the study of the emitted electron energy spectrum. Let us begin with a little kinematical exercise. Let a particle A decay into two other particles B and $C : A \rightarrow B + C$. The energy E of a particle of mass m and momentum $\vec{p}$ is given by: $E = \sqrt{m^2 + |\vec{p}|^2}$. In the rest frame of particle A, momentum conservation gives $\vec{p}_B + \vec{p}_C = 0$. Energy conservation gives $m_A = E_B + E_C$. It follows that

$$E_B = \frac{m_A^2 + m_B^2 - m_C^2}{2m_A} \; ; \; E_C = \frac{m_A^2 + m_C^2 - m_B^2}{2m_A}; \qquad \text{(A1.6)}$$

in other words, kinematics uniquely determines the energy of each one of the final particles. This reasoning, applied to reaction (A1.5) which was believed to describe β-decay, implies that the emitted electrons *ought to be mono-energetic.*

Indeed, the first experiments, between 1906 and 1914, were roughly in agreement with this result. These came mostly from Berlin, the chemistry lab of Otto Hahn (1879–1968, Chemistry Nobel Prize 1944) and Lise Meitner,[13] in which the electron energy was estimated by its penetration length. Poor precision allowed them to interpret the results as a series of mono-energetic rays and to attribute the absence of a single ray to impurities in the radioactive source.

The situation changed in 1914 thanks to James Chadwick, former student of Rutherford in Manchester (see Figure A1.7). Chadwick too arrived in Berlin to work with Geiger, Rutherford's former assistant in the α-scattering experiment. Chadwick used a magnetic field to measure the electron energy and a system of counters for their detection. The result was unambiguous: no unique ray. The spectrum appeared to be continuous[14] with the electron energy between zero and a few MeV. This result is in violent contradiction with equation (A1.6) which, as we

[13] A 'grande dame' of nuclear physics (1878–1968). Born in Vienna, she studied with Boltzmann and came to Berlin to work with Max Planck in 1907. Together with Otto Hahn, she played a very important role in all experiments on radioactive elements. The story goes that they had to work in an abandoned warehouse because the regulations, in effect up to 1909, did not allow women to access the main laboratory. She left Berlin in 1938, just in time to escape Nazi persecution and found shelter in Sweden.

[14] This experiment offers several fascinating aspects. Firstly the result, which, as we have noted, was against all established ideas. Secondly the novel experimental method. Finally, the absence of any apparent contact, let alone

Figure A1.7 The neutrino saga: James Chadwick (1891–1974, Nobel Prize 1935); Wolfgang Pauli (1900–58, Nobel Prize 1945); Enrico Fermi (1901–54, Nobel Prize 1938); Pauli : CERN Pauli archives; Fermi : Univ. Roma I, Amaldi archives.

explained, is a straightforward consequence of energy and momentum conservation. This was the first *energy crisis*.

The result was confirmed by an experiment in 1916, but apart from that, nothing happened until the beginning of the 1920s; there was a war going on. The following years were those of great confusion. Several experiments were reported among which we find the name of the young Ellis. We shall come back to him. Hahn and Meitner performed a new series of measurements, always insisting on their interpretation of mono-energetic rays. They proposed an ingenious theoretical scheme which was roughly the following. The decay produces electrons with energy given by formula (A1.6). But we should recall that the electrons

collaboration, between Hahn's team in chemistry and that of Geiger in physics. This is surprising since they were all in Berlin, they were studying the same decays and, even more so, they were all coming from Rutherford's lab. But what follows is also interesting. Chadwick found himself in Germany when the war started. He was arrested and interned at a prison camp in Ruhleben for the duration of the war. The surprising thing is that Chadwick found the means to continue his experiments in the camp, kept his correspondence with Rutherford, and continued to receive scientific publications as well as visits from German colleagues, such as Geiger and Otto Frisch. In the camp he met Charles Drummond Ellis, an officer cadet of the British army who was on holiday in Germany and was also interned in Ruhleben. Apparently, Chadwick's enthusiasm for science was such that Ellis, four years younger, decided to give up his plans for a military career and study physics. We shall have the occasion to present his work shortly.

were supposed to be among the nucleon's constituents. Therefore, an electron ejected from the interior of the nucleus, has a 'primary' energy given by (A1.6), but it loses part of this by rescattering before getting out. Thus, the continuous spectrum. Apparently unanswerable. Other explanations of a similar nature were also proposed.

At this point Ellis arrives. After the war he studied physics at Cambridge and joined Rutherford's team at Cavendish. He published some work on nuclear spectra, which showed already his great skill in experimental techniques, but his masterpiece is a revolutionary method for resolving experimentally the controversy on β-decays.

How can we measure the primary energy of an electron, if we only have access to its final energy? Ellis and William Alfred Wooster[15] found the answer and in doing so, established a new experimental technique, *calorimetry*.[16] Their reasoning was as follows. Experiments show a continuous spectrum for β-decay electrons. It is comprised between zero and a maximum value E_{max}. But we do not know whether it is the 'primary' energy, or that we observe after rescattering. Let us consider a radioactive substance. We can measure the number N of decays in a time interval T. If we manage to measure *all* the energy E_{tot} released during the time T, we can have the answer without knowing the path of each individual electron. Because, if Hahn and Meitner are right, we must find $E_{tot} = NE_{max}$, while, if Chadwick is right and the primary electrons are indeed released with a continuous spectrum, the result will be $E_{tot} = NE_{av}$, where E_{av} is *the average value* of the energy. Therefore, we need a calorimeter and not a spectrometer.

It took Ellis and Wooster two years to build and operate their calorimeter. The increase in temperature they measured was of order 10^{-3} degrees Celsius. The total energy deposited was proportional to the average value of order 0.34 MeV, far below the maximum value, which was larger than 1 MeV. This measurement put an end to all attempts to explain the spectrum by secondary effects. The year 1927 was that of the second, and most serious, *energy crisis*.

What was the reaction of the scientific community to this new measurement? Firstly, Lise Meitner declared she had felt a great shock.

[15] A student at Cambridge (1903–84), Wooster followed a scientific career in crystallography.

[16] Today calorimeters are essential components of any modern detector. The BEH boson identification at CERN was largely based on data produced by electromagnetic calorimeters.

As a good experimentalist, she repeated the experiment and confirmed the result. In her 1929 article she correctly acknowledges the work of Ellis and Wooster, but she proposes no explanation. Resolving the energy crisis was now left to the theorists.

No lesser man than Niels Bohr, in collaboration with Hendrik Anthony Kramers from Holland and John Clarke Slater from the USA, proposed in 1924 a theory in which all conservation laws, including that of energy, held only 'on average' and not for every individual process. These were the first years of quantum mechanics and Bohr was willing to upset any established principle. Einstein also envisaged such solutions for a while, but he soon gave up. Pauli remained very critical towards all these speculations. The energy crisis was the principal problem but, in fact, there were also others which increased the general confusion. The first spin measurements of various nuclei often gave results in apparent contradiction with angular momentum conservation. Pauli's exclusion principle, which explained atomic spectra very well, seemed to fail in nuclear physics. Niels Bohr, at a conference in London in 1930 ('The Faraday Conference') presented a good summary of all the problems encountered in nuclear physics. Today we know that they were due in large part to the nuclear model itself, which assumed the nuclei to be bound states of protons and electrons but, at that time, physicists had no solid basis for questioning this.

This brings us to December 1930. A conference was organised in Tübingen to discuss all problems related to what was termed at that time 'radioactivity'. Pauli was invited but decided not to attend. He sent a letter instead, dated December 4, written in his own light, inimitable style, in which he launches one of the most speculative, but also profound ideas. The letter starts with 'Dear radioactive ladies and gentlemen'. A little further down Pauli declares that he regrets not being able to participate in the conference, but his presence at a ball organised by the Italian Students Association in Zürich was indispensable. In the main part of the letter he writes '…I found a desperate way to solve the problem of statistics …as well as that of the continuous spectrum in β-decay. …the possibility of the existence inside the nucleus of neutral particles with spin 1/2. Thus the continuous spectrum could be explained by the hypothesis that one of these particles is emitted together with the electron in the decay …Your humble servant, W. Pauli'.

Pauli calls this new particle a *neutron* but the properties he attributes to it, in particular its mass,[17] are not those of the future neutron. It is also worth noticing that Pauli does not challenge the prevailing hypothesis according to which everything that comes out of a nucleus exists already inside it.

Pauli never published this proposal. He had probably submitted it to his friend Heisenberg, because the latter, in a letter dated December 1st, refers to 'your neutrons'. Pauli presented it to various conferences and in June 1931, the news about this new particle made headlines in the New York Times.

In January 1932, Chadwick discovered a new particle which he also called a 'neutron'. We thus had two particles with the same name. Fermi coined the term 'neutrino', meaning 'little neutron' in italian. The same year Heisenberg proposed the theory of isotopic spin, which we have presented in Chapter 3. Soon Chadwick's neutron became the accepted ingredient of nuclear matter, next to the proton, and nuclear physics took its present form. The following year Fermi arrived at a general synthesis with the theory of β-decay. It was the first theory formulated in the language of quantum field theory. It also established the idea that particles are created or annihilated by the interactions. Electrons and neutrinos don't 'come out' of a nucleus, they are created at the moment of decay.

Even if the controversy concerning possible violations of the conservation laws continued for some time,[18] the year 1933 marks the end of the first act of the neutrino saga. As we have noted already, it is a perfect example of the scientific process:

A new phenomenon is discovered and physicists launch a series of experiments which often yield incompatible results.
New experimental techniques are invented in order to obtain reliable measurements.
As far as possible, people try to interpret the results in the framework of existing theories.
When this is proven impossible, all avenues are explored, even the most speculative.

[17] Pauli estimates that the mass of this particle should be less than one hundredth of that of the proton.

[18] We find contributions by Bohr and even Dirac, as late as 1936.

At the end a new idea, often simple and elegant, emerges and imposes itself. The neutrino's existence ceased very soon to be a subject of controversy. In Fermi's theory the neutrino is a 'particle' like any other.

It is the end of the first act, but not the end of the story. Neutrinos revealed many more surprises throughout the twentieth century. We shall not tell the entire story here, we will just mark the most important milestones.

1. 1956: First direct observation of a neutrino. Frederick Reines and Clyde Lorrain Cowan[19] working in the Los Alamos nuclear reactor, observed a reaction produced by an (anti-)neutrino. In fact neutrinos have neither strong nor electromagnetic interactions. As a result, they interact with matter very weakly. For a neutrino with an energy of a few MeV the probability of having an interaction with a nucleus is extremely small. Hence the great difficulty in observing it. Nuclear reactors produce, by fission, a large number of unstable nuclei whose decay products contain neutrinos. Reines and Cowan succeeded in detecting the inverse of the reaction (A1.4): $\bar{\nu} + N_1 \rightarrow N_2 + e^+$, with the detection of the recoil of the nucleus N_2 and the positron, in coincidence.
2. 1957: Maurice Goldhaber measured the polarisation of a neutrino produced in β-decay. It was an incredible 'tour de force': measuring the polarisation of a particle we cannot see. One of the most beautiful experiments in the history of elementary particles.
3. 1962: The second neutrino. The arrival of large accelerators marked the era of a new source of neutrinos. They are those found among the decay products of the π mesons, predicted by Yukawa and presented in Chapter 4. Their main decay mode is $\pi \rightarrow \mu + \nu$, where μ is a new lepton, like the electron, but much heavier (see Section A1.5.2). So the question arose whether these neutrinos were the same as those produced in β-decay. In 1962 Leon Max Lederman, Melvin Schwartz, and Jack Steinberger, using the first neutrino beam built in the Brookhaven accelerator, showed that the answer was negative: there indeed exist two

[19] Clyde Lorrain Cowan Jr, born in 1919, died in 1974. For the first neutrino observation his collaborator Frederick Reines received the Nobel Prize in 1995.

distinct kinds of neutrinos.[20] Later we discovered a third.[21] We present these in the particle table at the end of this appendix. We have introduced the notation $\nu_{(e)}$, $\nu_{(\mu)}$, and $\nu_{(\tau)}$ in order to distinguish between them.

4. 1972: Neutral currents. In Section 6.4 we explained the importance of reactions like the elastic scattering of neutrinos on nuclei, which we called 'neutral current reactions'. These were first observed at CERN by the 'Gargamelle' collaboration.
5. 1998: First observation in Japan, by the 'Super-Kamiokande' collaboration of a strange phenomenon named 'neutrino oscillations.'[22] We have just explained that there exist three species of neutrinos. The aforementioned experiments showed that a neutrino produced in one of these species has a certain probability of manifesting itself later as belonging to another species. In other words, during their propagation, neutrinos oscillate among the three species. The first observation concerned solar neutrinos, but the phenomenon was confirmed for all neutrinos, irrespective of their origin. A beautiful illustration of the fundamental laws of quantum mechanics.

A1.5 The table of elementary particles

A1.5.1 The elementary particles in 1932: the world is simple

Before presenting the table of elementary particles known today, we start with the much simpler one of 1932. The year is not chosen randomly. 1932, the year the neutron was discovered, marks the beginning of elementary particle physics, the year in which our ideas on the structure of matter started to take their present form.

In 1932 we knew only one 'quantum of radiation', the photon, the quantum associated with an electromagnetic field. With regard to the constituents of matter, we knew, of course, the electron, we had just

[20] For this discovery Lederman, Schwartz, and Steinberger shared the 1988 Nobel Prize.

[21] The Large Electron–Positron (LEP) collider at CERN has shown that there are no further light neutrino species.

[22] For this observation Raymond Davis Jr. and Masatoshi Koshiba received the 2002 Nobel Prize. In fact, the 2015 Nobel Prize was attributed to Arthur B. McDonald and Takaaki Kajita, who studied this phenomenon further.

Table A1.2 The table of elementary particles in 1932.

Table of elementary particles	
Radiation quanta	
Photon (γ)	
Matter particles	
Leptons	Hadrons
ν_e, e^-	p, n

completed the doublet of nucleons with the discovery of the neutron, and two years earlier Pauli had postulated the existence of the neutrino in order to explain the electron spectra in β-decay (see Table A1.2).

This completes the presentation of the 1932 table of elementary particles. By looking again we can make the following remarks:

1. All matter particles have spin 1/2. The quantum of radiation has spin 1.
2. The algebraic sum of electric charges of all matter particles is equal to zero. At this stage this property appears to be a coincidence but, as explained in Section 6.4, it is important.
3. Every entry in Table A1.2 plays an important and well understood role in the structure of matter. This is obvious for the electron, the proton, and the neutron, which are the constituents of atoms and molecules. It is also obvious for the photon, which transmits electromagnetic interactions and allows for the formation of atoms. But it is also true for the neutrino, whose role is not visible at the classical level. In fact, neutrinos appear in neutron decay and make nuclei with a large neutron excess unstable. On the other hand, the same reaction is at the origin of stellar evolution and, indirectly, energy production in stars. The cycle of nuclear reactions in the interior of stars, such as that of the sun, starts with the fusion of two protons to give a deuteron, a neutron–proton bound state:

$$p + p \to d(p,n) + e^+ + \nu_{(e)}. \tag{A1.7}$$

In addition, the great transparency of matter to neutrino radiation, makes the latter essentially the only means of a massive star radiating away energy and lowering its temperature. Thus, neutrinos too, have a decisive role in the structure of the Universe.

A1.5.2 The elementary particles today

The disorder

This simple picture with a small number of elementary particles did not last for long. In 1937 a new particle was discovered in cosmic rays, which, after a few episodes, turned out to be a new lepton, called a *muon* (μ). Later, in 1962, it was established that it is accompanied by its own neutrino and it carries a new quantum number. In order to distinguish among the various neutrinos, we have put the lepton with which each one is associated as an index. Thus we write $\nu_{(e)}$, $\nu_{(\mu)}$, etc. The muon is 200 times heavier than the electron but otherwise, the doublet $(\nu_{(\mu)}, \mu)$ seems to have the same properties as the more familiar $(\nu_{(e)}, e)$. In retrospect, muon discovery marked a turning point in our concepts of elementary particle physics: it is the first particle whose specific role in the structure of matter still remains unknown.[23]

After the Second World War the proliferation of 'elementary' particles increased at a fast pace. Looking for the mediator of nuclear forces, whose existence was predicted in 1935 by Hideki Yukawa, physicists discovered a large number of hadrons. The first was precisely Yukawa's π meson[24] which was discovered in cosmic rays in 1947. With the operation of the large accelerators the discovery rate increased exponentially and today's Particle Data Book has more than one hundred entries. Even more importantly, all simple rules deduced from Table A1.2 were violated: among the new hadrons some had integer and some half-integer spin and the distinction between matter particles and radiation quanta was impossible. As for the role of each one of those particles in the structure of the world, physicists did not even dare ask the question. Complete disorder appeared to reign.

[23] It seems that it was I. Rabi who, after discovery of the muon, asked the question: 'Who ordered that?'

[24] From the Greek word μέσον, which means 'middle'. In fact, the first mesons to be discovered had masses intermediate between that of the nucleons and of the electrons.

The quarks

By the early 1960s we had accumulated a considerable number of hadrons and their properties had been sufficiently studied for physicists to be able to attempt to apply some order to this chaos.[25] Past experience had taught us that the enormous variety of atoms and molecules which surround us result from a surprisingly small number of constituents. It was thus natural to try to interpret all these hadrons as bound states of a few, more 'elementary' particles. There had been several attempts in this direction, but the model which was finally confirmed by experiment is the one proposed independently by Murray Gell-Mann and George Zweig in 1964. Its basic postulate is that all hadrons are made out of spin 1/2 building blocks, *the quarks*.[26] In 1964 we thought there were three types of quarks, today we know there are six. In Table A1.3 we write them as (u, d, c, s, t, b). Baryons are formed out of three quarks and mesons out of a quark–antiquark pair. For example, the proton is a bound state of two quarks u and one quark d, the neutron of one quark u and two quarks d. Table 6.1 contains more examples of hadron composition in terms of quarks. These examples show that the electric charge of the quark u equals 2/3 and that of the quark d is −1/3. Therefore, the 'elementary charge' in the hadronic world equals 1/3 times the electron charge. Similarly, the other four quarks, c, s, t, and b, form hadrons which are produced in our accelerators and are unstable. We can find them in the Particle Data Book, which is updated regularly. We shall not need them in this book.

A last remark concerning the number of quarks. We are in fact obliged to complicate this simple picture of six quarks, sole constituents of all hadrons. We have seen that quarks have spin 1/2, they are therefore fermions. In Section A1.3 we noticed that identical fermions have the strange property of anti-symmetry under the exchange of any pair among them. Let us consider the example of a hadron called N^{*++},

[25] In a review article in 1957 on elementary particles, Murray Gell-Mann and Edward P. Rosenbaum wrote: '... At present, our level of understanding is about that of Mendeleyev, who discovered only that certain regularities in the properties of the elements existed. What we aim for is the kind of understanding achieved by Pauli, whose exclusion principle showed *why* these regularities were there, and by the inventors of quantum mechanics, who made possible exact and detailed predictions about atomic systems ...'

[26] This is the name given by Gell-Mann. Zweig called them 'aces'.

which was discovered in the 1950s. It has a spin equal to 3/2 and an electric charge equal to 2. It is formed as a bound state of three quarks *u*. Its wave function must be anti-symmetric under the exchange of any pair of these. On the other hand we can convince ourselves of the following: (i) the spins of the three quarks must be parallel in order to build a total spin equal to 3/2, therefore the spin part of the wave function is symmetric; (ii) the space part must also be symmetric, because all theoretical computations show that the minimum energy bound state is obtained when the wave function is symmetric under the exchange $\vec{x} \leftrightarrow \vec{y}$. We conclude that the anti-symmetry cannot come from either the position or the spin of the quarks and therefore we must have another variable associated with them. This reasoning led to the introduction of the concept of *colour*. This assumes that each one of the six quarks (u, d, c, s, t, b) may exist in three different types. We call them *colours*, but we want to stress that there is no relation with the ordinary sense of the word. It simply implies that we should attribute to each quark an extra index u_i, d_i, ..., where i takes the values 1, 2, and 3. It is with respect to this index that the N^{*++} wave function is anti-symmetric. This sounds like an artifact, just another way of saying that we have in fact 18 quarks. Although the counting is correct, the physical interpretation is not. We must supplement the model with an assumption, fully confirmed by experiment, that every hadron contains all three colours in equal proportions and matter appears, in fact, colourless.

When this idea was first proposed in 1964[27] it did not raise a great deal of enthusiasm. It was gradually imposed in the 1970s when it was found that its predictions were verified by experiment and it could provide the basis for a fundamental theory of strong interactions, the one presented in Section 6.3. A basic difficulty, not easy to explain, was the fact that states with a single colour never appear in nature.

This brings us to the other strange property of quarks, which we also presented in Section 6.3. The presence of quarks inside hadrons has been verified experimentally. The first results came from the scattering of high energy electrons off nucleons. It is the analogue of Rutherford's experiment which established the existence of the atomic nucleus. Here

[27] The first proposal, in a slightly different form, was made by O. W. Greenberg. It was followed a year later by a similar one by M. Y. Han and Y. Nambu for integer charge quarks.

Table A1.3 The table of elementary particles today. This table reflects our present ideas on the structure of matter. Quarks and gluons do not appear as physical particles and the graviton has not been observed.

Table of elementary particles		
	Radiation quanta	
Strong interactions		Eight gluons
Electromagnetic interactions		Photon (γ)
Weak interactions		Bosons W^+, W^-, Z^0
Gravitational interactions		Graviton (?)
	Matter particles	
	Leptons	Quarks
1st family	$\nu_{(e)}, e^-$	$u_i, d_i, i = 1, 2, 3$
2nd family	$\nu_{(\mu)}, \mu^-$	$c_i, s_i, i = 1, 2, 3$
3rd family	$\nu_{(\tau)}, \tau^-$	$t_i, b_i, i = 1, 2, 3$
	BEH boson	

the experiments showed the presence of hard grains inside protons and neutrons. Nevertheless, we have never succeeded in breaking a nucleon and taking out a quark. As we explain in Section 6.3, we have good reasons, experimental, numerical, and theoretical, to believe that this is due to a property of the interaction which binds the quarks inside the hadrons. The force between the quarks increases with the distance; weak at short distances, it becomes very strong, possibly infinite, when the separation becomes macroscopic and tends to infinity. We called this property *confinement* and its rigorous derivation from first principles is always a challenge for theorists. We present it schematically in Section 6.3.

The table today

With the arrival of quarks the table of elementary particles again took a rather simple form (see Table A1.3).

The table presents the elementary particles, or rather those assumed to be, in three sectors: the *radiation quanta*, the *matter particles*, and, finally, the famous *BEH boson*.

1. *The radiation quanta.* We have seen the role of the photon as the messenger of electromagnetic interactions. In Chapter 4 we saw that

this concept of messenger is generalised to all interactions. The particles which appear in this first sector of Table A1.3, are precisely the messengers of the four interactions presented in Section A1.2. The geometrical theory we have developed in this book determines uniquely their number and their properties. There are *eight gluons* for strong interactions; *the photon* for electromagnetic interactions; *three intermediate bosons*, denoted by W^+, W^-, and Z^0, for weak interactions, and *the graviton* for gravitational interactions. All these particles, with the exception of the graviton, have been identified experimentally. They are all bosons. The gluons, the photon, and the intermediate bosons of the weak interactions have spin equal to 1. We shall call them collectively *gauge bosons*. The spin of the graviton is expected to be 2.

2. *The matter particles*. Here we find the particles which, according to our present understanding, are the elementary constituents of matter. They are spin 1/2 fermions and in Section A1.4 we saw that the fermionic character is essential for the formation of nuclei and atoms. In the table these particles are classified into three groups which we call *families*. Let us start by looking at the first one: it contains the electron e and its neutrino $\nu_{(e)}$, together with the pair of quarks u and d. The index i which takes three values denotes the three colours. This first family resembles the matter particles of Table A1.2 which we considered elementary in 1932, with the quarks u and d replacing the nucleons, proton, and neutron. All macroscopic matter is made out of constituents belonging to this first family.
   The other two families appear as copies of the first. Each one contains a doublet of leptons, with one charged lepton, μ and τ, and one neutrino, as well as a doublet of quarks in three colours. These heavy quarks are also confined and produce new hadrons. The leptons μ and τ, as well as the new hadrons, are all unstable and decay into particles of the first family.
3. *The BEH boson*. This is the subject of this book. It is electrically neutral and its spin is equal to zero. According to our present understanding, it is an independent component in the world of elementary particles. In Chapter 7 we presented some theoretical speculations attempting to guess a new physics beyond that of the Standard Theory. In such a framework this boson could be linked with the other particles of Table A1.3.

We end with the three rules which we formulated after the 1932 table. The first two remain valid. Matter particles have spin 1/2 and all radiation quanta which have been identified so far, have spin 1. The sum of electric charges in each family is still equal to zero. It is worth noticing that, in order to verify this rule, we need the hypothesis of the three colours for the quarks. Indeed the electric charge of each lepton is −1. The quarks of each family contribute $2/3 - 1/3 = 1/3$ and we must multiply this result by 3 in order to obtain the value +1. However, the third rule is no longer valid. If we understand very well the role of the first family particles in the structure of matter, we know of no good reason for the existence of the other two. Why do we have three families, since just one would have been sufficient? Why does nature reproduce what seems to be three copies of the same thing? Such questions show the limits of our understanding of the fundamental physical laws.

APPENDIX 2

From Sophus Lie to Élie Cartan

Introduction of the *group* concept is relatively recent in mathematics. It appears for the first time, although only implicitly, in the work of Leonhard Euler in the eighteenth century. The term 'group' was first used by Évariste Galois in around 1830. As is often the case in mathematics, this initial concept was enriched with time, becoming more and more abstract. Since it underlies all ideas on symmetries that we use in this book, we shall present in this appendix a few elementary notions. Even in a very simplified form, they remain quite technical and require some familiarity with the language used in mathematics. Some additional simple concepts will be explained as we go along. The aim of this appendix is to help the reader who wishes to follow the reasoning which led to the physical theories we have presented in this book, but it is by no means necessary to understand the results. We shall mainly talk about the work of two great mathematicians, the Norwegian Marius Sophus Lie and the Frenchman Élie Joseph Cartan (see Figure A2.1).[1]

[1] Marius Sophus Lie, one of the great figures of nineteenth-century mathematics, was born in Nordfjordeide, Norway in 1842. He wanted to follow a military career, but he was forced to give up because of poor vision. He studied at the University of Christiania (today Oslo) but it was only in 1867, at the age of 25, that he decided to major in mathematics. He travelled a lot to meet the great mathematicians of his time. It seems that it was in Paris, in contact with the French mathematician Camille Jordan, that he discovered the importance of group theory in geometry. At the out break of the 1870 war he decided to leave Paris on foot to go to Italy, but he was arrested as a German spy and imprisoned in Fontainebleau. He was liberated only after the intervention of Jean Gaston Darboux and returned to Norway. He occupied the chairs of Mathematics at the Universities of Christiania and Leipzig and, between 1888 and 1893, published a three volume monumental work under the title *The Theory of Group Transformations.* His reasoning follows a strong geometrical intuition, not always appreciated by his fellow mathematicians at the time. In 1874, he married Anna Birch, a great-niece of Niels Henrik Abel, another famous Norwegian mathematician

Figure A2.1 Marius Sophus Lie (1842–99); Élie Joseph Cartan (1869–1951).

Rather than introducing the abstract and precise definition of the general group concept, we shall limit the discussion to a particular case,

who died in 1829 at the age of 27. Lie had worked in editing Abel's collected works. In poor health himself, he died in Norway in 1899.

Élie Joseph Cartan, born in 1869 at Dolomieu, is considered to be one of the greatest French mathematicians from the turn of the nineteenth century. He entered the École Normale Supeure (class 1888), where he followed the courses of famous mathematicians like Goursat, Hermite, Darboux, and Poincaré. Between 1892 and 1894 he studied at the Collège de France with a scholarship from the Pécot Foundation. It was during this period that he corresponded with Sophus Lie. In 1894 he published his first results on the complete classification of all finite dimensional Lie algebras, one of the most remarkable mathematical results of the end of the nineteenth century. He was 25 years old. In 1910, after a study of the representations of the three-dimensional rotation group, he introduced the notion of *spinors*. Notice that in physics, the concept of *spin* was only introduced in 1925 by the Dutchmen Samuel Goudsmit and George Uhlenbeck. Cartan made many other important contributions which turned out to be interesting for physics, such as the classification of symmetric spaces and a reformulation of differential geometry for general relativity. An outstanding teacher, he taught mathematics at the University, the École Normale Supeure, and the École de Physique-Chimie. He is considered to be one of the main founders of the modern school of French mathematics. He died in Paris in 1951.

which happens to be the only one we have discussed in this book. This is the concept of transformations of the coordinate system in a given space.[2]

Definition: Let $\mathcal{G}$ be a set of transformations which act on the coordinate system of some space. We denote a particular transformation by g. We shall say that $\mathcal{G}$ forms a group if the following conditions are fulfilled.

1. *If we apply to the coordinate system two successive transformations g_1 and g_2 belonging to $\mathcal{G}$, the result is equivalent to applying to the coordinate system a transformation g_3 which also belongs to $\mathcal{G}$. We write this composition property as $g_1 g_2 \rightarrow g_3$.*
2. *$\mathcal{G}$ contains the trivial transformation g_0 which leaves the coordinate system unchanged. We call it 'identity'.*
3. *If $\mathcal{G}$ contains the transformation g, it contains also the inverse transformation, denoted g^{-1}, such that the successive application of g and g^{-1}, taken at any order, brings the coordinate system to its initial state. We write this property as $gg^{-1} = g^{-1}g \rightarrow g_0$.*

It is obvious that all the transformations we consider in this book, translations, rotations, inversions, etc. form groups. We can distinguish between *finite groups*, which contain only a finite number of transformations, and *infinite groups* which contain an infinite number. For example, space inversion is a group with only two elements, to wit the identity $\vec{x} \rightarrow \vec{x}$ and the proper inversion $\vec{x} \rightarrow -\vec{x}$. The same with time inversion. The symmetry groups of regular crystals are also examples of finite groups. Contrary to this, rotations or translations are infinite groups.

Another notion which is intuitively obvious is that of a *sub-group.* The rotations around one axis is a sub-group of the three-dimensional rotation group.

In this book we have also used a property of group transformations which is familiar from rotations. Consider the rotations around an axis. Each one is characterised by an angle. It is easy to write the composition law: the result of two successive rotations, one with angle θ_1 and a second one with θ_2, is equivalent to a rotation with angle $\theta_3 = \theta_1 + \theta_2$. It is also obvious that we obtain the same result if we change the order and apply the rotation with angle θ_2 first and θ_1 second. We say that

[2] In fact, this is not far from Lie's geometrical ideas. A mathematician would say today that we study some particular group representations.

these transformations *commute* and we call the group *commutative*. We also call it *abelian*, in honour of the Norwegian mathematician Niels Henrik Abel (1802–29).

We have also encountered *non-commutative*, also called *non-abelian*, groups. The simplest example is the group of rotations in a three-dimensional space. It is easy to check that the result of the composition of a rotation around the x axis and one around the y axis, depends on the order in which we perform them. From the mathematical point of view, non-abelian groups are more interesting because they have a richer structure. The Yang–Mills gauge theories we have presented in this book are based precisely on non-abelian groups.

Were we to give free rein to our imagination, we could define a large variety of groups, including groups of transformations, because the definition we have given is very general. It was the great merit of Sophus Lie that he chose a particular class among infinite groups[3] which is large enough to cover most of the interesting cases and precise enough to allow for detailed study. We shall not follow his introduction, which today represents only a historical interest, but our approach, which follows a physicist's motivations, respects his own geometrical spirit.

We have seen that the elements of the transformation groups we have been considering depend on one or several parameters. Rotations on angles, translations on vectors, etc. Lie understood that an essential notion which will characterise the properties of these groups is that of *continuity*, therefore we can restrict ourselves to the study of groups *whose elements depend continuously on the parameters.* Even if the notion of continuity sounds intuitively obvious, its precise definition is quite technical and we are not going to present it here. On the other hand it is clear that this choice allows us to apply to groups a good part of the theory of continuous functions, and we can convince ourselves that these groups, which we call *Lie groups*, have rich and multiple properties which make them interesting to both mathematicians and physicists.[4]

[3] The study and classification of finite groups turned out to be a very interesting chapter of modern mathematics and it was completed only during the second half of the twentieth century. We shall not have occasion to use it in this book.

[4] Rather than engaging in lengthy and abstract definitions, it may be instructive to give an example of an infinite group of transformations which is *not* a Lie group. Let us take our familiar example of the group of rotations around an axis. Its elements depend on the rotation angle θ which we may decide to

A consequence of this property of continuity, which was already noticed by Lie, is that, for a given group, we can study its structure by limiting ourselves to small transformations in the neighbourhood of the identity element. For example, if we want to study the group of rotations, it is sufficient to consider rotations with infinitesimal angles. The property of continuity will allow us as a next step to reconstruct large angle rotations by a succession of 'small' rotations. At first sight this remark sounds rather trivial, but Lie understood that the small transformations in the neighbourhood of the identity give rise to a new structure which encodes all the essential properties of the group. Today we call this structure *Lie algebra*, but Lie called it an *infinitesimal group*. Once more, we shall not give the precise definition but we shall show its importance with some particular examples.

Consider again the example of the group of rotations around an axis. We have seen already that the composition law is given by the summation over the angles: $\theta_3 = \theta_1 + \theta_2$. θ is an angle, therefore it takes values between $-\pi$ and $+\pi$. We can visualise the transformations as the motion on a circle. If we limit ourselves to infinitesimal angles, θ also measures the length of the arc on the circle, therefore the same law also describes the group of translations along an axis. We have just seen the interest of this analysis: two groups, which are globally very different, motion on a circle from $-\pi$ to $+\pi$ and on a line from $-\infty$ to $+\infty$, are *locally* the same, or, using the technical term, *they have the same Lie algebra*.

We can object that this example is trivial and one could have guessed the result by inspection. This objection is valid, but we can give more sophisticated examples for which the tools developed for the study of Lie algebras will be essential. To do so, we need the work of Cartan.

We have already introduced the number of independent transformations we can perform in the space we are considering. This number will

measure, for example, in degrees. Let us now consider a particular sub-group formed by the rotations whose angle, expressed in degrees, is given by a rational number (a number which we can write as a ratio of two integers). It is obviously a group because the sum of two rational numbers is a rational number and the other two conditions are also fulfilled. It is not a Lie group because we do not have the property of continuity; between two rational numbers there is an infinity of irrational numbers. So this group is not covered by Lie's definition and it is also clear that it is not particularly interesting.

play an important role in the discussion that follows, so we shall give it a name: *the dimension of the Lie algebra*. We have seen that for rotations in a three-dimensional space, this number is equal to three. It corresponds to the three independent rotations we have learnt in geometry. A warning: this equality between the number of space dimensions and that of the independent rotations is a numerical accident, valid only in three dimensions. For example, if we consider a plane (two dimensions), we have only one rotation around the axis perpendicular to the plane. With a bit of imagination we can convince ourselves that the number of independent rotations we can perform in a four-dimensional space equals six.[5]

Up to now we have considered only spaces with real coordinates. But already in Section 3.3, when we were discussing isotopic spin symmetry, we remarked that in an internal symmetry space the basic vectors of the coordinate system may be the particle fields which, like the wave functions, can take complex values. Therefore we must define transformations in such spaces. We did it implicitly in Section 6.3 when we talked about quantum chromodynamics as the theory invariant under transformations which mix the quark fields with different colours. It is obvious that the further we go in this direction, the more abstract the discussion becomes and it loses a simple geometrical significance. Cartan's results allow us to put a definite order to this labyrinth of spaces and transformations.

In 1894 Cartan obtained the complete classification of all finite-dimensional Lie algebras,[6] a result of capital importance for both physics and mathematics. He showed that there exist five classes of Lie algebras, four of them form infinite series and the fifth contains five exceptional cases. For each algebra he gave also its dimension, i.e. the number of independent transformations. His notation is not the one we use in physics, but we can summarise his results as follows.

[5] The counting for any dimension is easy: we know that in a plane we have only one possible rotation. So, let us count planes rather than axes. A plane is defined by two axes, therefore we must compute the independent pairs of axes. In n dimensions each axis can pair with $n-1$ of the remaining ones. This gives us $n(n-1)$. But this counts every pair twice, so the final result is $n(n-1)/2$. For $n = 2$ we get 1, for $n = 3$ we get 3, for $n = 4$ we get 6, etc.

[6] If we want to be more precise we must add '...for semi-simple groups on complex numbers', but these properties will not be needed in our discussion.

1. For the rotations in an n-dimensional real space we write the group as $O(n)$, where O is an abbreviation for 'Orthogonal'.[7] As we have seen, the dimension of the corresponding algebra equals $n(n-1)/2$.
2. For the transformations in an n-dimensional complex space we write the group as $SU(n)$. 'S' stands for 'special' and means that we do not consider a group of transformations which simultaneously change the phase of all vectors. 'U' stands for 'Unitary' which means that the transformations do not change the modulus of the vectors.[8] The dimension of the algebra equals $n^2 - 1$.[9]

We can search in this list the groups we have used for the Standard Theory.

(i) The group of rotations around a single axis is not in the Cartan list. It is an abelian group with a one-dimensional algebra which mathematicians consider to be trivial. We denote this by $U(1)$.

(ii) In Cartan's classification the algebra of the group $O(3)$, i.e. the rotations in a three-dimensional real space, is identical to that of $SU(2)$. We used this result implicitly when we said that the Heisenberg group of isotopic spin, whose transformations act on the two complex vectors given by the proton and neutron fields, is equivalent to the group of three-dimensional real rotations; see Chapter 3. The dimension of the algebra equals 3, hence the three weak interaction gauge bosons W^+, W^-, and Z^0.

[7] In Cartan's notation the algebras of these groups form two of his four infinite series, which he denotes as B_n $n \geq 2$ for the groups $O(2n+1)$ and D_n $n \geq 4$ for the groups $O(2n)$.

[8] Cartan notes this series as A_n $n \geq 1$ for the group $SU(n+1)$.

[9] The counting is also relatively simple. A general transformation applied to a set of n complex vectors depends on n^2 complex numbers, which give $2n^2$ real numbers. It is convenient to write them as a tableau with n lines and n columns. We call these tableaux *matrices*. We must now count the constraints implied by the unitarity condition, i.e. the conservation of vector modulus. In the matrix notation this condition can be written as $U(n)U^*(n) = \mathbf{1}$, where $\mathbf{1}$ is the unit matrix whose only non-zero elements are on the diagonal and are all equal to 1. $U^*(n)$ represents the generalisation of the concept of complex conjugation for matrices. This condition gives n^2 constraints. A further constraint comes from not counting the transformation of a common change of phase for all vectors. The result is $n^2 - 1$.

(iii) The algebra of the group $SU(3)$, which is the group of unitary transformations acting on the three-dimensional complex space of the three coloured quark fields, has dimension $3^2 - 1 = 8$. Indeed, the number of gluons, the gauge bosons of quantum chromodynamics, equals 8.

Index